CW00827964

THE RECENT GEOLOGY

OF

CORNWALL.

BY

W. A. E. USSHER, ESQ., F.G.S.,

OF H. M. GEOLOGICAL SURVEY.

[*Extracted from the* GEOLOGICAL MAGAZINE, from January to July, 1879.]

CONTENTS.

PART I.

PART II.

PART III.

PART IV.

PART V.

THE

POST-TERTIARY GEOLOGY OF CORNWALL.

PART I. HISTORICAL.

TO ascertain the most recent movement to which a country has been subjected, and by careful comparison with the past to discover what insensible changes are now progressing, is of the utmost importance in approaching its Quaternary History.

By a recourse to such occasional observations as have been recorded by historians or monkish chroniclers, gleaned perhaps in few cases from actual investigation, and exaggerated, no doubt, in an age delighting in the marvellous some information may be gained, but when we consider that these notes were made rather for the gratification of the curious than with a view to ascertain their causes or to forecast their effects, and that the facts of one century may become the legends of the next, it behoves us to sift the evidence, retaining only such bare and unvarnished statements as by incidental mention and simple relation appear most worthy of credence, especially when the accounts are corroborated by independent writers

It has ever been the characteristic of the ignorant and uninquiring peasantry to ascribe the occurrence of great boulders of rock dissimilar to any in the neighbourhood the fantastic shape, so frequently effected by weathering in rocks of unequal durability and such-like remarkable objects, to the agency of fabulous beings endowed with enormous strength and gigantic proportions, and so names are given to phenomena of unusual occurrence, and are retained by a less credulous posterity even when the legends which suggested them have almost entirely passed away. Many such names are to be met with in Cornwall

Again, traditions of a more extensive coast-line, of lands now swept away, have been handed down, doubtless magnifying the extent of the ancient land, as the account passed through succeeding generations

Our familiarity with the causes producing such phenomena as earthquakes, comets, eclipses, and the like, however seldom some of them have been experienced in a lifetime, renders the observations of the present age more accurate and less liable to exaggeration than those of preceding centuries, when anything of infrequent occurrence in the experienced operations of nature was regarded as cataclysmal, resulting from direct interposition in an unvarying state of things. The rapid advance and more general cultivation of scientific research, no longer fettered by ignorance and superstition, embraces in an ever-extending chain of cause and effect phenomena which our ancestors regarded as supernatural.

It is however curious to note how some amongst the ancients by the acuteness of their perceptions, grasped an occasional scientific truth which has been corroborated in the present day. Thus, it is remarkable that Ovid, Pythagoras, Pliny, and Aristotle should have believed the sea to be less changeable than the land.[1] Strabo, in opposing the opinions of Eratosthenes and Xanthus as to the cause of shells being found at great elevations and distances from the sea, says: "It is not because the lands covered by the seas were originally at different altitudes that the waters have arisen or subsided or receded from some parts and inundated others. But the reason is that the same land is sometimes raised up or depressed so that it either overflows or returns to its own place again. We must therefore ascribe the cause to the ground, either to that ground which is under the sea or to that which becomes flooded by it, but rather to that which lies beneath the sea, for this is more movable."

The historical evidence may be classified under three heads:—

Firstly, accounts of unusual disturbances of the sea by contemporary observers.

Secondly, records of disastrous inundations preserved in old chronicles.

Thirdly, traditions of the Lyonesse and probable references of the ancient geographers and historians to the Scilly Isles.

Fourthly, the insulation of St. Michael's Mount and the identification of Ictis.

PART 1.—*Contemporary Observations.*

These have been taken exclusively from papers by Mr. Edmonds. In Edin. New Phil. Journ. he mentions an influx and reflux of the sea, varying from three to above five feet, in Mounts Bay, at five P.M., on March 23rd, 1847; the double movement taking from fifteen to twenty minutes. During the most part of the day the water, from the mouth of the Catwater to within Sutton Pool, at Plymouth, was constantly agitated by flux and reflux.

In Falmouth Harbour, and on the shores of the Scilly Isles, similar oscillations took place, whilst in St. Ives Bay nothing unusual was remarked.

[1] Stoddart, Proc. Brist. Nat. Soc. for 1870, vol. v. p. 43.

At Newlyn four fluxes and refluxes of the sea occurred in an hour and a half. In the shallow water between Marazion and Penzance no agitation was perceptible. The limits of the disturbance, so far as observed, were from Mousehole on the west to Porthleven on the east, a distance of ten miles.

On October 30th of the same year, at five P.M., a rise of the sea, coming from the south-west, and reaching five feet, took place at Penzance.

Three similar fluxes and refluxes occurred at Plymouth in forty minutes.

Four whirlwinds, accompanied by shocks, passed through the parish of St. Just, on December 12th, 1846.

The same writer[1] mentions an earthquake felt over 100 miles, from the Scilly Isles through Cornwall as far as Plymouth, in July, 1757.

A disturbance of the sea took place in Mounts Bay at four hours and a quarter after the great earthquake at Lisbon in 1755, when the sea suddenly rose to the height of six feet at St. Michael's Mount coming in from the S.E., and to eight feet at Penzance Pier, coming in from the S.E. and S.S.E. At Newlyn Pier and Mousehole the sea coming in from the south rose and fell ten feet. Toward the decline of the commotion, the sea was found to be running at seven miles an hour in Guavas Lake.

If the observation recorded in the following extract be not magnified in transmission from the original observer, it shows the care necessary in ascribing the occurrence of some isolated pebbles and boulders above the reach of the highest spring tides to changes in the relation of sea and land. "I have been informed by two descendants of an eye-witness that at Lamorna Cove, which is on the south-east part of Mounts Bay, the sea on this occasion rushed suddenly towards the shore in vast waves with such impetuosity that large rounded blocks of granite from below low-water mark were swept along like pebbles, and many were deposited far above high-water mark. One of seven or eight tons weight was rolled to and fro several feet above high-tide level."

Whether the size of the boulders be exaggerated or not, it is evident that the disturbance described was sufficiently powerful to shift large stones from the existing beach to a point about the average height of the Cornish raised beaches above high-water mark, even allowing for an exaggeration of five feet in the height to which the large boulder was said to be moved. At Polkerris, near the Par estuary, 'Raised Beach" has been engraved on the map, apparently on the strength of the occurrence of isolated quartz pebbles amid sandy debris on a small promontory some twenty feet above the adjacent beach, which is composed of exactly similar quartz pebbles. This phenomenon is much more likely to have been produced by exceptional gales, or such disturbances as have been described, than to be the relics of a raised beach, the lighter

[1] T. R. G. S. Corn., vol. vii. p. 164, etc.

materials of which had been dissipated by spray and rain, for the raised beaches are usually too much consolidated to allow of such facile dissipation.

In February, 1759, Mr Edmonds records a slight shock felt at Liskeard for fifteen minutes, accompanied by blood-red rays

In March, 1761, on the day of the second earthquake at Lisbon the sea advanced and retreated five times four hours and a quarter after ebb-tide, at five P.M., in Mounts Bay, rising six feet at Penzance and Newlyn, and four feet at St. Michael's Mount At the Scilly Isles the agitation continued for more than two hours.

In July, 1761, fluxes and refluxes occurred in Mounts Bay, and at Falmouth, Fowey, and Plymouth

In 1789 fluxes and refluxes of the sea were observed at Penzance and St. Michael's Mount. Earth shocks were felt on December 30th, 1832

In 1836 a slight disturbance of the earth was felt in the parishes of Budock and Stithians.

On October 20th, 1837, a slight shock is said to have been felt in the Scilly Isles.

On February 17th, 1842, an earthquake was felt between the hours of eight and nine A.M., from Manaccan on the south to St Cubert on the north, a distance of twenty-five miles, and from Falmouth on the east to St. Hilary on the west, a distance of eighteen miles

On July 5th, 1843, the sea was much agitated within Porthleven Harbour. Three hundred yards from the north shore of the harbour nothing unusual was observed At one P.M. the sea rushed in for fifty yards, reaching a height of four or five feet at Marazion At Penzance an agitation accompanied by strange currents was observed.

The effects of the disturbances above cited are eminently transient, except in abnormal shifting of detritus to higher levels, but when we find that within the short space of a century Cornwall has felt the spent force of earthquakes propagated from distant centres of internal or eruptive motion, the probability of similar disturbances emanating from much nearer sources, and productive of considerable if not permanent effects is at once suggested Whilst the record of such cataclysms in early historic or mediæval times would refer to their disastrous effects, want of knowledge and observation leaving the causes unknown, the recent prehistoric geological period conceals them in an impenetrable veil.

PART 2.—*Records of Disastrous Inundations.*

I quote the following from Mr Peacock's book (On Vast Sinkings of Land, etc.) :—

p. 116 "Dr Barham quotes from the Saxon Chronicle, particulars of the inundation of Nov. 11th, 1099, and of another on the same authority in 1014. This year (1014) on Michaelmas Eve,

Sept. 28th, came the great sea-flood which spread over this land, and ran up as far as it never did before overwhelming many towns and an innumerable multitude of people."

p. 115. An account of a destructive inundation 13 years after the Domesday Survey, by Florence of Worcester; "On the 3rd day of the Nones of November 1099, the sea came out upon the shore, and buried towns and men very many, and oxen and sheep innumerable." From the Saxon Chronicle for that year, "On St Martin's mass day, the 11th Nov., sprung up so much of the sea-flood, and so myckle harm did, as no man minded that it ever afore did, and there was the ilk day a new moon." "Whence," says Mr. Peacock, "the catastrophes cannot be referred to the great height of the tide, for the highest spring-tides do not occur until several tides after the new moon, and the 11th of November is several weeks after the equinox."

p. 138. Mr. Peacock accounts for Geoffery of Monmouth's omission of the mention of the inundations of 1014 and 1099, on the ground that the chroniclers very often omitted to record the actual disappearances of lands.

In p. 140 he quotes from Mr. Pengelly's paper on the Antiquity of Man in the South-West of England: "Leland (1533-1540) says, 'Ther hath been much land devourid betwixt Pensandes and Mouse-hole. Ther is an old legend . . . a Tounlet in this Part (now defaced and) lying under the water.'"

In p. 141 he gives a reference to Mounts Bay from Magna Britannia published anonymously in 1722 (vol. i. p. 303). "'Tis a tradition among the people here, that the ocean breaking in violently, drowned that part of the country which now is the Bay." Mr. Peacock disposes of the idea that the catastrophes of 1014 and 1099 might have been the result of similar movements to those "which occurred on the South Coast of England in 1817, 1824, and 1859, at a considerable distance of time from either equinox," on account of the unprecedented harm done by them, and the inadequacy of such high tides as those mentioned to produce commensurate effects.

Notwithstanding, I am inclined to differ from Mr. Peacock in this conclusion for the following reasons:—

Firstly. Such traditional accounts as those of Leland and the Magna Britannia, and the statement of Vice-Admiral Thevenard in Mem. relatifs à la Marine (A.D. 1800), "La submersion du terrain . . . et de la pointe ouest de l'Angleterre IX" (commencement of ninth century), quoted by Mr. Peacock in p. 88 of his book, must be laid out of the question.

Secondly. All statements made by writers who lived long after the occurrences they describe must be accepted with reservation, as they may have been derived from the contemporary record of the occurrence, and cannot, therefore, be said to furnish additional evidence. Thus with Florence of Worcester, who wrote in the thirteenth century.

Thirdly. Taking the Saxon Chronicle as the only direct contemporary account of the inundations of the eleventh century, one would like to know whether the descriptions there given were penned by an eye-witness of the catastrophe, or inserted from rumours which would doubtless have magnified the disaster ere they reached the chronicler.

Fourthly. Admitting Mr. Peacock's reason for the omission of remarkable events here and there by the chroniclers generally, I cannot see their particular application to Geoffery of Monmouth, who flourished in the twelfth century, and would therefore have less excuse for omitting to mention events, which had been witnessed by the generations immediately preceding him, than Florence of Worcester, who lived more than three centuries after they had occurred. For these reasons I am disinclined to believe in sudden elevations or depressions of land, and to consider that, owing to some such disturbances as I have quoted from Mr. Edmonds, though perhaps of greater magnitude, lives may have been lost and lands devastated by the influx of waves propagated by earthquake shocks, and by seasons of unprecedented flood. That the effects produced would be partial or transient, whilst the story of the disaster for which men could assign no cause would be magnified as it passed from the eye-witnesses of the catastrophe to their descendants and finally, with many interpolations and distortions, live as a local tradition with perhaps very little of its original significance remaining.

PART 3.—*Traditions of the Lyonesse, &c.*

The following information is chiefly extracted from Mr. Peacock's book:—

"It is said that in Camden's time the inhabitants of Cornwall were of opinion that the Land's End did once extend further to the west, which the seamen positively conclude from the rubbish they draw up, and that the land there drowned by the incursions of the sea was called Lionesse. That a place within the Seven Stones is called by the Cornish people Tregva (*i.e.* a dwelling), and that windows and other such stuff have been brought up from the bottom there with fish-hooks, for it is the best place for fishing. That at the time of inundation supposed Trevelyan swam from thence (at least 15 nautical miles to the nearest part of the mainland) and in memory thereof bears Gules, an horse Argent issuing out of the sea proper." (*Vide* Note A.)

"If the Lyonesse country really existed in Ptolemy's time (A.D. 117 to 161), it cannot have extended as far westwards as is shown on the map in the Churchman's Magazine (for July, 1863, p. 39), from Land's End and Lizard Point to and comprising the Scilly Isles. Because Strabo, who flourished at least a century before Ptolemy, quoting Posidonius, who was still older, mentions those islands as then existing under the name of Cassiterides (book iii. cap. ii. § 9), and that they were ten in number (*Ibid.* cap. v. § 11)."

"Dr. Paris, in his 'Guide to Mounts Bay and the Land's End,'

p. 91, mentions Camden's tradition of the Lyonesse (the Silurian Lyonois), said to have contained 140 parish churches, all of which were swept away by the ocean." He says further that the Scilly Isles are now 140 in number, though only six are inhabited.

Camden (Britannia, edit. 1722) says, "The Scilly Isles are called by Antoninus, Sigdeles; by Sulpitius Severus (died A.D. 420), Sillinæ; by Solinus, Silures; by Dionysius Alexandrinus, Hesperides; by Festus Avienus (latter part of fourth century), Ostrymnides; by several Greek writers, including Diodorus, and by Pliny the Elder, Cassiterides."[1]

Dr. Borlase,[2] in a letter to the Rev. J. Birch on the Scilly Isles, says that the present inhabitants are new comers, having no connexion with the old race, as all the antiquities found in the islands belong to the rudest Druidic times.

In isles now uninhabited and not used for pasturage, rude stone pillars, erect circles of stone, kistvaens, innumerable rock basins, and tolmens,[3] are found, whilst the small islands, tenements, and creeks, are called by British names.

Within the three years previous to 1753, he states that the advance of the sea in the Scilly Isles has been very considerable; this advance being, in his opinion, due to subsidence for the following reasons: Strabo's opinion as to their number (*vide suprà*) and as to one only being desert and uninhabited; the fact that the Isle of Scilly, which gives its name to the group, is now a high barren rock, a furlong across, with cliffs to which only sea-birds can obtain access.

The flats which stretch from one island to another are plain evidences of a former union between many now distinct islands. The flats between the islands of Trescan, Brehar, and Sampson, are left quite dry at a spring-tide low-water, when walls and ruins have frequently been seen through the shifting sands, covered by 10 to 20 feet of water at high tide. As these foundations were probably at one time six feet at least above high-water mark, the advance of the sea by denuding action alone would be insufficient to account for their present position, "ten feet below high-water." Whence he considers that "a subsidence amounting to 16 feet at least has taken place, which caused the desertion of the islands by their terrified aboriginal inhabitants. These original inhabitants carried on a trade in tin with the Phœnicians, Greeks, and Romans" (for this opinion he cites Diodorus Siculus, lib. v. cap. ii. and Strabo, Geog. lib. iii.). "Whilst only one inconsiderable vein of tin occurs in Tresco Island, and that betrays no sign of ancient working, nor are any old workings now visible sufficient to have maintained a trade in tin." He says further, "But though there are no evidences to be depended

[1] Peacock, p. 109.

[2] Phil. Trans. for 1753, vol. 48, p. 326.

[3] Tolmens.—Oval or spheroidal rocks, when resting on two others, with a cavity between, are called by Dr. Borlase tolmens (stones with holes), and are supposed by him to have been rock deities (Carne on the Scilly Isles).—T.R.G.S. Corn. vol. vii. p. 144.

on of any ancient connexion of the Land's End and Scilly, yet tha the cause of that inundation which destroyed much of these island might reach also to the Cornish shores, is extremely probable, there being several evidences of a like subsidence of the lands in Mount Bay."

Dr. Borlase, in his Natural History of Cornwall,[1] says, "The supply of tin from Gades and Spain being too small to supply the vast trade as far as India, they must have got it to the east of the Damnonii."

The Chaldeans and Arabians call tin by a name similar to the Greek κασσίτερος The Scilly Isles were called Cassiterides long before the Greeks knew of their position, for Herodotus (B C 400) says, Οὔτε νήσους οἶδα Κασσιτερίδας ἐούσας, ἐκ τῶν ὁ κασσίτερος ἡμῖν φοιτᾷ (book iii. cap. 115).

Solinus calls them Insulæ Silurum or Insula Silura, perhaps in mistake for islands off the Welsh coast.

Tacitus[2] says the Silures were opposite to Spain, which would point to the Scilly Isles. It is probable that the Phœnicians regarded West Cornwall as an island, and one of the Cassiterides, as the Scilly Isles alone would have been totally insufficient to afford the supply.

"Ortelius,[3] therefore, not without reason, makes the Cassiterides to include, not only the Scilly Isles, but also Devonshire and Cornwall."

"Tin was also anciently found in Lusitania and Gallicia."[4]

Mr. H. Boase[5] quotes Carew[6] as follows:—"The encroaching sea hath ravined from it the whole country of Lionnesse, together with divers other parcels of no little circuit, and that such a Lionnesse there was, these proofs are yet remaining. The space between the Land's End and the Isles of Scilly, being about 30 miles, to this day retaineth that name, in Cornish, Lethowsow, and carrieth an equal depth of 40 or 60 fathoms, save that about midway there lieth a rock which at low water discovereth its head. They term it the Gulf, suiting thereby the other name of Scilla. Fishermen also casting their hooks thereabouts, have drawn up pieces of doors and windows." After touching on Dr. Borlase's views, Mr Boase[7] proceeds to say, "The arguments adduced by our old historians in proof of the tradition, refute themselves. In the first place, the sea is no shallower between the Land's End and Scilly, than at equal distances from land, on other parts of the coast; and the midway gulf or Wolf-rock, happens not to be in that channel at all, but considerably to the south of it, and as to the stories of fishing up pieces of doors and windows, and seeing tops of buildings, etc., had all the buildings, doors, and windows of Cornwall, been placed there, the first tempest would have swept them all away, as pebbles before a torrent. The truth is, that no such relics were ever discovered, or could have remained for discovery, in that boisterous channel of the Atlantic Ocean."

[1] p. 29. [2] Ib. p. 30. [3] 1527–1593. [4] Ib. p. 29.
[5] T. R.G.S. Corn. vol. ii. pp. 130, 131. [6] Carew, p. 3. [7] Op. cit. p. 132.

With the above opinion I entirely agree, for the very mention of windows dredged up is sufficient to refute any testimony of an historical connexion of the Land's End with the Scilly Isles based upon it. Except as fragments of wreck, it is impossible to conceive the occurrence of such material in the places specified.

(Peacock p. 140.) The tradition of the loss of area on the West of Land's End is thus mentioned by Harrison (An Historical Description of the Island of Britaine, by W. Harrison, prefixed to Hollingshed's Chronicles, 1586, vol i lib. iii. ch 10, p 397): "A remarkable corroboration of Ptolemy's positions of the promontories Belerium and Ocrinum,"[1] as Mr. Peacock thinks, "It doth apeere yet by good record, that whereas now there is a great distance betweene the Syllan Isles and point of the Land's End, there was of late years, to speke of scarslie a brooke or drain of one fadam water betweene them, if so much, as these evidences appeereth and are yet to be seene in the hands of the lord and chiefe owner of those Isles."

Dr. Paris and Mr Carne[2] considered that St. Just in the Land's End district might have been meant by the word Cassiterides, owing to the traces of tin in the Scilly Isles being insufficient to justify that appellation. Mr. Carne,[3] speaking of Piper's Hole, in Tresco Island, as a supposed adit of the ancient tin works, objected that as it is above high-water, it is just such a site as would be selected now. He further considers that, if any mines had ever been productive in the Scilly Isles, some traces of diluvial tin ore would even now be found from time to time in the low-lying tracts in St Mary's, and on the south-eastern side of Tresco

Mr. Peacock[4] quotes Diodorus Siculus as follows:—"Far beyond Lusitania (Portugal) very much tin is dug out of the islands in the ocean nearest to Iberia (Spain), which from the tin are named Cassiterides."

D P Alexandrinus, who flourished in the time of Augustus, says in his Geography, line 599, etc.: "But beyond the Sacred Promontory (Cape St. Vincent), which they affirm is the extremity of Europe, in the islands Hesperides, where the source of tin is, the rich children of the illustrious Iberi dwell." Mr. Peacock thinks that the Scilly Isles are here alluded to under the name Hesperides.

Strabo has told us that Publius Crassus saw that the metals were dug out at a little depth in the Cassiterides (book ii. cap. v. § 15); this was about 57 B.C.

Strabo further describes the Cassiterides as "islands in the high seas just under the same latitude as Britain, northward and opposite to the Artabri."[5]

[1] Peacock, p. 109.

[2] Mr Carne (T R G S, Corn. vol. ii. p. 354) says, "It is exceedingly probable that the western extremity of England, of which St. Just forms a prominent part, constituted the principal portion of what was formerly known under the name of the Cassiterides"

[3] T R G S. Corn. vol. vii. p. 153.

[4] Peacock, p. 106.

[5] Peacock, p 107

PART 4.—*St. Michael's Mount*

The best description of St. Michael's Mount, as it now exists, that I can find, is by Mr Pengelly,[1] as follows: "The Mount is an isolated mass of granite measuring about five furlongs in perimeter at its base. At high-water it plunges abruptly into the sea, except on the northern or landward side, where the granite comes in contact with the slate, into which it sends veins and dykes, as may be well seen on each side of the harbour. Here there is a small plain occupied by a village, adjacent to which is the harbour, built in 1726-7, and, as Mr Johns, the harbour-master, has been good enough to write me, capable of receiving ships of 500 tons burthen" Its situation is described as follows: 'The distance between the nearest point of Marazion Cliff and spring-tide high water mark on the Mount is 1680 feet. A tidal isthmus (Hogus) of highly inclined Devonian slate and associated rocks, in most cases covered with a thin layer of gravel or sand, is at spring-tide high-water, in still weather, 12 feet below, and at low-water 6 feet above the sea-level. This ridge is dry in fine weather from four to five hours every tide but occasionally during storms and neap tides it is not passable for two or three days.'

"St Michael's Mount[2] was named in Cornish, as Carew informs us 'Caraclowse in Cowse, in English, the hoare rock in the wood: which now is at every flood encompassed by the sea, and yet at some low ebbs, roots of mighty trees are descried in the sands about it.' . . . . . Florence of Worcester expressly asserts that it was formerly five or six miles from the sea and enclosed with a very thick wood, and therefore called in British, Carreg lug en Kug, 'Le Hore Rok in the wodd.''

The above is said to have been corrected by Florence of Worcester in a letter to William of Worcester, 1478.[3]

Mr. Peacock[4] thinks that we need not go back further than the time of the Domesday Book for the origin of the Cornish name of St Michael's Mount, 'Carreg coedh yn clos," *i.e.* "Rock of the wood in the enclosure' as William Camden (1550-1623) "proves that the Cornish language had not become quite extinct even so lately as his time"

"Dr Gibson,[5] the editor of Camden's Britannia, says that St. Michael's Mount is called Carreg Cowse in Clowse. Carreg is, doubtless, the origin of the English word crag, and cowse is said to mean cana, white, and clowse obviously means a close or enclosure"

"Mr. Metivier says that St Michael's Mount was 'Carreg Coed yn Clôs,' rock of the wood in the enclosure"

Mr Peacock[6] says that "the earliest period at which the Saxon

[1] Journ Roy Inst. Corn. for 1873, p 12.
[2] T R G S Corn., vol. ii p 134.
[3] Pengelly on Submerged Forests in Torbay.
[4] Peacock, p 110.
[5] Ibid p 89
[6] p 111

name Mychel Stop, or Michael's Step, could have been given to the Mount, was after the landing of Hengist and Horsa in 449."

The Mount received its present name in 1085, from the Monastery of St Michael, of which it then became an appanage; before that time it was called Dinsol[1]

"In Milner's Gallery of Nature, p. 387, it is stated that in the time of Edward the Confessor, 1044, the rock of St Michael's Mount was the site of a monastery described as being near the sea, 'juxta mare' (interpreted by Dr Barham, 'by the sea')."

"The ancient designation," says Mr Pengelly, "betokens a change in the geography of the district—a change, not only within the human period, but since Cornwall was occupied by a people who spoke the language which was tardily supplanted by the Anglo-Saxon."

Mr Pengelly refers the name 'Hogus," now applied to the rocky ledge between Marazion and the Mount, to an old Scandinavian derivation, meaning "a rock in or near a wood adjacent to water, and used for sacrificial purposes."

Mr Peacock[2] takes exception to this determination on the ground that Hogus (in Guernsey hougue, French hogue, neo-Latin hoga) sometimes denotes a quarriable knoll, of which he gives examples. From this Mr Peacock infers that the term Hogus only carries us to the middle ages, and not to the time of Diodorus.

Mr Peacock[3] quotes Diodorus Siculus (about 44 B.C.) as follows. "They who inhabit the promontory Belerium are exceedingly hospitable, and on account of the merchants being their guests are civilized by custom in their mode of life. They procure the tin by ingeniously working the earth producing it, which, being rocky, has earthy veins, in which working a passage and melting (the ore) they extract [the tin]. Forging it into masses like Astragals, they carry it into an Island situate before Britain, called Ictis. For the middle space being dried by the ebb they carry the tin into this (island) in abundance in carts. (But a certain peculiar thing happens concerning the neighbouring islands lying in the middle (μεταξυ) between Europe and Britain, for at full sea they appear to be islands, but by the reciprocation of the ebb of the sea, and a large space being dried, they appear peninsulas.) Hence the merchants buy [the tin] from the inhabitants and export it into Gaul."

Taking μεταξυ to mean 'in the middle," Mr Peacock considers that the Northern Channel Islands were alluded to in the above passages, being of opinion that the Northern Channel Islands were then only insulated at high-water, and that they are called neighbouring islands to distinguish them from the more remote islands in the Bay of Biscay.

Mr. Pengelly[4] observes that, according to Leland, St Michael's Mount in 1533 was no larger than at present, that William of Worcester's estimation of its distance from the mainland differs but little from its present site that "Bishop Lacy's encouragement to the Faithful in 1425 to complete a causeway between Marazion and

[1] Ibid p. 112.
[2] Peacock p. 107.
[3] Peacock, p. 86.
[4] Journ. Roy. Inst. Corn. for 1873, p. 181.

the Mount, for the protection of life and shipping, denotes that the exposure was as great as in our day; and as the Confessor's Charter in 1044 describes the Mount as 'juxta mare,'[1] next or by the sea, it may be safely concluded that the insulation of the Mount had taken place more than eight centuries ago."

After a passing allusion to other competitors for the Ictis of Diodorus, he says, "It is perhaps worthy of remark, that those who have studied the Geology of Cornwall, espoused the cause of the Mount; while those who fail to do so, appear to have come to the question with their minds imbued with a belief in William of Worcester's statement, that there were 140 parish churches submerged between the Mount and Scilly, and accordingly hold that the submergence took place not only since the time of Diodorus, but since the introduction of the parochial system into Cornwall."

Mr. Pengelly quotes Sir George Cornewall Lewis (An Historical Survey of the Astronomy of the Ancients) as follows: "Timæus mentions an island of Mictis within six days' sail of Britain which produced tin, and to which the natives of Britain sailed in coracles." He regarded Mictis and Ictis as variations of Vectis.

From Mr. Pengelly's statement that the Mount 1900 years ago possessed a harbour, Mr. Peacock dissents on the ground that "if the coast had remained unaltered ever since Diodorus's time, the Roman tin-transporting ships need not by any means have been confined to St. Michael's Mount as a harbour, because, as the Rev. W. Borlase[2] well observes, Guavas Lake is the principal anchoring place." Whence he considers that the chief export of tin could not have taken place from St. Michael's Mount, and does not favour the belief in its identification as the Ictis of Diodorus. He says further:[3] "The ancient block of tin which was dredged up about 1823 in Falmouth Harbour (Lyell's Principles of Geology, 1867, p. 451), if we suppose it to have been dropped during its transit to the Isle of Ictis, would seem to place Ictis opposite Falmouth harbour, and therefore twenty miles east of St. Michael's Mount."

Mr. Pengelly, in a lecture at the Royal Institution,[4] says, "The Mount is by no means a solitary rock of its kind. Within seventy miles east of it there are certainly four that actually are or probably were, within the last 1900 years, precisely similar though slightly larger islands—Looe Island, St. Nicholas Island, the Mewstone, and Borough Island."

Mr. Peacock cherishes the idea that the Mounts Bay forest was submerged in the historic period, and is sufficient confirmation of the "tradition of these parts that St. Michael's Mount, now enclosed half a mile with the sea, when the tide is in, stood formerly in a wood."

He quotes the following note from Carew (1602):[5] "Tradition tells us that in former ages the Mount was part of the insular continent in Britain, and disjoined from it by an inundation or encroachment of the sea, some earthquake or terrestrial concussion."

[1] "Sanctum Michaelum qui est juxta mare."

[2] Phil. Trans. vol. 48.

[3] Peacock, p. 118.

[4] Quoted by Peacock, p. 139.

[5] p. 140.

"If," says Mr Peacock,[1] "the storm of 1099 and Dr Borlase's submersion[2] in the ninth century be true, St Michael's Mount cannot have been the ancient isle of Ictis, because must we not suppose that the Mount only became an island at one of these submersions." Mr Peacock strengthens his position by the following quotation[3] from page 2 of the Domesday Book: 'The land of Michael . . there are two hides which never paid the Danish tax (nunquam geldaverunt). The land is eight caracutes."

The hide is generally supposed to be equal to 120 acres.[4] Sir H. Ellis says that the measure of a hide varied in different places at different times. "The caracute was as much arable land as could be managed with one plough and the beasts belonging thereto in a year, having meadow, pasture, and houses for the householders and cattle belonging to it."

Taking the smallest estimate of a "hide" from the five different measures of it in the reigns of Richard I., Edward I., and Edward II., which vary from 60 to 180 acres, Mr Peacock says that eight caracutes would have amounted to 490 acres, whilst[5] the present dimensions of the Mount, measured from the Ordnance Map, "are found to average 22 × 14 chains; the area therefore is 30·8 acres,[6] and it is quite clear that, so far from there being eight caracutes of arable land, there can hardly be a single acre capable of being ploughed, because the ground is too steep and rocky."

Mr Peacock[7] believes that at the date of this description in the Domesday Book (in the year 1086), St Michael's Mount was not an island, for the following reasons. Firstly, because neither the Domesday Book nor the Saxon name Michael Stop give any reason for such a conclusion. Secondly, because it is the custom in the Domesday Book, "when a place is an island, to call it so." Of this he gives examples. Thirdly, on account of its then containing at least eight times as much land as at present.

Of the several remaining competitors for the Ictis of Diodorus, Mr. Peacock disposes as follows:—

As the Scilly Isles do not lie between Europe and Britain, and as there is a 43-fathom sounding between them and the Land's End, none of them would answer to the description of Ictis.

As to the Isle of Portland or the Isle of Wight, so accurate an observer as Diodorus would not have failed in distinguishing their position definitely as "near the south coast of Britain, nor are there any grounds for the supposition that the relations of either locality to the mainland were different in Diodorus's time from the present."

With respect to the claims of Mont St. Michel, he considers that the space between it and the Continent was the Forest of Scisy and not sea until seven centuries and a half after Diodorus's time.

[1] Peacock, p. 88.

[2] Dr. Borlase was inclined to refer the submersion of St Michael's Wood to the inundation of the year 830, mentioned in Irish Annals. Mr. Whitaker ascribed it to that mentioned by the Saxon Chronicle and Florence of Worcester as occurring in 1099. *Vide* T. R. G. S. Corn., vol. ii. p. 139.

[3] Peacock, p. 137. [4] Ib. p. 113.

[5] Ib. p. 135. [6] Ib. p. 114. [7] Ib. pp. 112, 113.

As alternatives, Mr. Peacock proposes the Wolf Rock (which would be opposite Britain if a westerly and north-westerly extension of the Cornish coast be conceded), the Seven Stones, or some island now totally lost He considers, however, that the identification of Ictis is "both impossible and unimportant"

Mr Claypole[1] gives an estimate of the uniform rate of depression of Mounts Bay on assuming the identity of St Michael's Mount with the Ictis of Diodorus and the Ocrinum of Ptolemy He says "It must then have been an island as now at high-water only. In the time of Diodorus the isthmus must have been below high-water mark. So depression must be restricted to limits allowing the isthmus to have been below the upper limit of 20-foot tide, 1800 years ago, and above its lower limit now so that it would not have exceeded 6 feet, therefore the rate of depression would be 4 inches per century, which would be 6 feet in 12,600 years."

Mr Pengelly,[2] commenting on the evidence furnished by the caverns of Devon, gives the following general note, which may not be out of place here "In order to obtain the whole, we must add to this part the time represented by the lodgement of the blue Forest Clay of Devon or the tin ground of Cornwall, to this again must be added the period in which the forests grew; to this a further addition must be made of the time during which the entire country was carried down at least 70 feet vertically by a subsidence so slow and tranquil and uniform that it nowhere throughout the area of Western Europe and the British Islands disturbed the horizontality of the old forest soil; and finally we must also add the time which has elapsed since—a time which of itself, thanks to the description of St Michael's Mount by Diodorus Siculus, we know certainly exceeded 2000 years, and which the volume of the stratified deposits overlying the forests, as well as the amplitude of the existing foreshore, warrants our believing exceeded it by a very large amount."

Conclusion.—If the word Cassiterides, in the writings of Strabo, Posidonius, and Diodorus, refers to the Scilly Isles, and if they have also been mentioned by Dionysius Alexandrinus under the name of Hesperides, the quotations from these authors would imply the following consequences.

First,—That tin must have been obtained in the Scilly Isles as they then existed

Secondly,—That, as no productive tin veins or signs of old workings are found on these islands, such workings must have been carried on in districts now submerged, at a time when the number of the islands (allowing a considerable margin on the score of insignificance in Strabo's account) was much less than at present, and when the flats between the islands of Tresco, St Mary, and St Martin (as may reasonably be inferred from Dr. Borlase's description), were dry land at high-water and above the level of spring-tides.

Thirdly,—That the Channel Islands were not insulated in Diodorus'

[1] Proc Brist Nat Soc 1870, vol v p. 35.
[2] Journ Royal Instit Corn for 1873

time; for, if they were, he would hardly have alluded to the Scilly Isles as nearest to Iberia. This accords with Mr. Peacock's views as to their more recent insulation.

Fourthly,—From Alexandrinus' account, must we not suppose that the inhabitants of the islands were a colony from Spain in his time, and either supplanted the original inhabitants alluded to by Dr. Borlase, or were themselves succeeded by a British race, addicted to Druidic rites?

Notwithstanding, I am inclined to think that the word "Cassiterides" was indiscriminately used for the Scilly Isles and Land's End District,[1] owing to the imperfect navigation of those early days of naval commerce.

Diodorus's description of the inhabitants and mineral wealth of Belerium would apply rather to a district of that name than to an individual promontory, and it does not seem improbable that the name of one of its most important headlands should be indiscriminately applied to the whole stanniferous district of the Land's End. If, as Mr. Peacock supposes, the Northern Channel Islands are spoken of by Diodorus as neighbouring islands with reference to Ictis, one can scarcely agree with him in disposing of the claims of the Isle of Wight to the appellation of Ictis on the ground of the accuracy of that historian's descriptions. If the name Vectis[2] applied exclusively to the Isle of Wight, Pliny's mention of it as lying between Ireland and Britain would prevent one from putting too much faith in the latitudes and longitudes of ancient geographers. (Vide Note B.)

To revert to more recent records. As the description of St. Michael's Mount in the Domesday Book is so indefinite, and, from the nature of the record, rather applicable to the lands belonging thereto than to the geographical position of the Mount itself, there appears to be little reason why the eight carucates mentioned in the passage should not be regarded as arable lands on the adjacent mainland belonging to the monastery.

The submergence of the Mounts Bay forest seems to have occurred considerably anterior to any inundation on record, for the following reasons.

First,—Mr. Carne[3] mentions the extension of the old forest ground seaward, traced to a depth of from twenty to thirty feet below spring-tide level.

Secondly,—There is every reason to conclude, with Mr. Carne, that the forest bed met with in a pit at Huel Darlington mine, under 12 feet of marine sediment, four feet of peat, and eight feet of river wash, is continuous with the forest bed on the beach.

Thirdly,—Whilst the entombment of the forests in marine sediments indicates subsiding movements, the peat and overlying gravel in

[1] T. R. G. S. Corn. vol. vii. p. 153.

[2] (Peacock, p. 183.) Pliny, Nat. Hist. lib. iv. § 30. "Sunt autem xl. Orcades modicis inter se discretæ spatiis. Septem Acmodæ, et xxx. Hebrides, et inter Hiberniam ac Britanniam, Mona, Monapia, Ricina, Vectis," etc.

[3] T. R. G. S. Corn. vol. vi. p. 230, etc.

Marazion Marsh, and the present positions of rock platforms slightly higher than spring-tide at high-water, and of estuarine deposits, seem to point to a slight subsequent elevation, not yet counteracted. The changes which took place after the submersion of the old forest ground can hardly have been comprised in eight centuries, and were more probably operating during a period of more than 2000 years. A belief in the pre-historic[1] submergence of the Mounts Bay forest is by no means contrary to the identification of St. Michael's Mount with the Ictis of Diodorus, for although the land may have been at a slightly lower level in the time of Diodorus than at present, the rapid disappearance of thirty-six acres of pasturage from the West Green sand banks[2] since Charles the Second's time, mentioned by Dr. Boase (T. R. G. S. Corn. vol. iii. p. 131) leaves one free to infer that prior to that time the bank was of still greater extent, so that its eastward portion may have facilitated the passage to the Mount by affording a ridge or causeway of sand covering the rocky isthmus and passable in most conditions of the tides. The Wolf Rock and the Seven Stones can scarcely be regarded as possible competitors for the Ictis of Diodorus; their admission would entail a subsidence of at least 200 feet within 2000 years, as the former is seven miles to the south-west of Guethenbras Point (Land's End district), with intervening depths of from twenty-one to thirty-eight fathoms; and the latter are fourteen miles west from the Land's End, with depths of thirty-two to forty fathoms between them and the Longships. St. Michael's Mount appears better to accord with the description of Diodorus than any other island on the Cornish coast on account of, firstly, its vicinity to tin-producing districts; secondly, the facility with which carts laden with the ore could have reached it either on the supposition of an elevation of a few feet, or allowing the extension of the sand-bank from the mainland or the existence of a sand spit concealing the isthmus.

APPENDIX.

NOTE A.—A rock near the Land's End bears the name of "the Armed Knight." Though this appellation may have been bestowed on it through a fancied resemblance in outline, the existence of the tradition respecting Trevelyan's adventure appears to furnish a more likely reason for the name.

NOTE B.—In Speed's Map of Cornwall, 1610, no dependence can be placed upon the latitudes, as may be seen by placing a tracing of a reduced Ordnance Map of the same scale (about 1 inch to 4 miles) over it, when the Land's End district will be found to occupy entirely different positions scarcely overlapping in any place, and the shape of the Lizard district to be quite dissimilar.

Another map without date, but probably as old as Speed's, was shown to me by Mr. Parfitt, of the Devon and Exeter Literary Institute; the same discrepancies were visible in it.

Now when we find discrepancies of latitude equal to 10′, and the shapes of promontories entirely misrepresented in maps of their own country produced by geographers 300 years ago, how can we expect to find even as great accuracy in the geographical descriptions of Roman or Greek historians, more especially when relating to coasts with which they must at best have been very slightly acquainted?

[1] As far as Britain is concerned.

[2] The banks are now only two or three acres in extent.

ERRATA.—p. 7, line 31, *for* "10 to 20" *read* "10 to 12," p. 8, line 12, *for* "400," *read* "410."

PART II.

POST-TERTIARY GEOLOGY OF CORNWALL

THE materials for a classification of the later Pleistocene deposits of Cornwall are so voluminous that it was found impossible to embody them in a single paper. Having elsewhere attempted a general classification with such notices of the deposits as seemed necessary to show the grounds whereon it was based, I purpose in the following paper to complete the notices of deposits. As an apology for the amount of compilation thus rendered necessary, I must plead the object of the papers, viz. to place in one view all that has been written on the subject, as references alone would entail more time and trouble in looking up than many readers would be disposed to concede.

The paper is divided into the following sections.—

1. Oldest superficial deposits; 2. (*a*) Boulder Gravels, (*b*) Raised Beaches, and (*c*) "Head." 3. Submerged Forests and Stream-Tin Gravels. 4. Recent Marine and Blown Sands.

1. Oldest Superficial Deposits

From their isolated positions, and evident relations to an entirely different surface configuration, the gravels of Crousa Down and Crowan and the sands and clays of St. Agnes, must be regarded as the earliest traces of superficial deposits as yet observed in Cornwall.

Gravels of Crousa Down and Crowan.—On Crousa Downs, Lizard District, a patch of rounded and subangular quartz gravel "occupies an area of about half a square mile at a height of about 360 feet above the sea" (Report on Geol. Corn. and Dev. p. 396).

The Rev. E. Budge (Trans. R. G. S. Corn. vol. vi. pp. 1 and 91) describes the deposits, generally, as extended layers of fine yellow gravel, with a quantity of quartz pebbles, exposed in pits 10 to 12 feet deep in places, near the road leading to Coverack. The character of the sections is given thus—Black peaty soil containing small angular quartz stones about 6 inches thick, upon layers of fine and very coarse gravel alternating in no very determinate order, containing quartz pebbles of very irregular form, some as large as a man's head, but for the most part not exceeding 2 to 3 inches in length. The Crousa Down gravel rests on Diallage rocks.

A similar occurrence was noticed by Mr. Tyack (62nd Ann. R. Geol. Soc. Corn. p. 176, etc.) at Blue Pool in Crowan. The pebbles covered an area of 800 yards from north to south, and 500 from E. to W.

They are scattered over the surface are well worn, and vary in size from large boulders to the dimensions of hazel nuts The gravel is 400 feet above the sea it rests on yellow clay As at Crousa Down the quartz is such as would be furnished by veins in the Killas, the pebbles of schorl being very few, and the occasional granite fragments angular, yet the Killas districts near Crowan are at a much lower elevation than the granite on which the pebbles are found.

These quartz gravels appear to have been derived from quartziferous Killas, either by direct transport of aqueous agencies sufficiently protracted in their operation to allow of the comminution of the slaty matter, or indirectly by the disintegration and redeposition of a quartz conglomerate rock of Palæozoic age. Referring to the derivation of the Crousa Down Gravel, the Rev. E. Budge illustrates the prevalence of quartz veins in the Killas to the north by citing the occurrence of masses of quartz in the slates near Nare Point, whence they can be traced for some miles along the line of strike; and of a quartz vein 10 feet wide on the south-west of Carne, in St. Anthony parish.

Between the Loo Pool and Marazion, on the top of the cliffs, near Trewavas Head, small flint and quartz pebbles occur in the soil, and do not appear to extend more than a few paces inland. As their height above the sea and the adjacent configuration preclude the possibility of their being the relics of a raised beach, I am forced to conclude that they are either traces of gravels somewhat similarly situated to those of Crousa Down and Crowan, or that during exceptionally severe gales some of the smaller pebbles of the beach below had been from time to time carried upward in the spray and landed on the top of the cliff.

Deposits of St. Agnes.—(Report, etc., p. 258). De la Beche was disposed to regard the sands and clays which nearly encircle the higher parts of St. Agnes Beacon as "the remnant of some supracretaceous deposit." "They occur at an elevation of between 300 and 400 feet above the level of the sea, resting upon the slates of the hill, and partly also on a small portion of the granite rock which there occurs; the granitic rock and slates being traversed by several tin lodes." "This isolated deposit has not hitherto been found to contain organic remains, with the exception of some traces of plants that have the appearance of Fucoids."

The following sections are given by De la Beche (Report, p. 259), Hawkins (Trans. Roy. Geol. Soc. Corn., vol. iv. p. 135, etc.), Henwood (*op. cit.* vol. v.), respectively—on the North-east of the Beacon. Numbers affixed for reference:

(1)	Head of rubble from hill above or Cobb	3ft.	0in.
	Yellow sand	21ft.	0in.
	Brownish sand with numerous planes dipping at 40° (apparently bedding)	11ft.	0in.
	Light-coloured mining clay	2ft.	0in.
	Blue clay	9ft.	0in.
	Yellow sand	4ft.	0in.
	White sand	4ft.	0in.
	Yellow sand	3ft.	0in.
	Pebbles resting upon an uneven surface of slate—thickness variable		

(2) Near Trevaunance—

Yellow Cobb with Killas rubble		
Fire clay	2ft	0in.
Clay and sand	3ft	0in
Fine white gritty sand—depth not ascertained		

(3) Half a mile from the Beacon—

Surface 383 feet above high water.

Clay	2ft	0in
Yellow sand. 7 feet below the surface	8ft	0in

The overburden not mentioned would seem to be 5 feet thick

The following sections given by Messrs Kitto and Davies lie toward the North-east of the Beacon (Trans. R G Soc Corn., vol. ix.).

(4) Near the outer margin of the deposit—

Head	5ft	0in.
Yellow sand	2ft	0in
Red sand	2ft	0in.
White sand	4ft	0in

Pebbles in sand not gone through.

Sections near the above on N W and S W

(5) On N W.—

Head	6ft	0in.
Clay with a few pebbles	3ft	0in
Sand	12ft	0in
Sandstone	2ft	0in
Sand with pebbles not gone through		

(6) On S W.—Very sandy overburden, with numerous quartz pebbles from the size of a marble to that of a walnut, beneath which clay only is raised varying from 6 to 12 feet in thickness.

On the inner margin of the deposit to the west of the above, "mining operations in 1865 exposed a cliff facing North, 16 feet in height, and 15 feet below the surface," nearly perpendicular, "smoothed and polished" and worn "into caves and hollows." An adit cut through to the sand on the other side proved this to have been in all probability a projection from a main cliff face, which old miners state to occur facing eastward for some distance to the southward, and to be worn into numerous hollows. The sand in this part of the deposit "contained very large pebbles and boulders and angular" stones.

Compare the section given by De la Beche (*op cit*) with the following by Hawkins on the North side of the Beacon (*op cit*), and by Henwood, locality not specified (*op. cit*).

(7) Yellow Cobb with rubble of Killas stones	2ft	6in
Brown sand with sedimentary divisions dipping S at 45°	9ft	0in.
White Clay		4in to 5in
Brown and bluish-grey clay (with a slight admixture of carbonaceous matter)	9ft or 10ft	0in
Gritty sand	6ft or 7ft	0in
(8) Loose stones and earth, up to	6ft or 8ft	0in
Pink, yellowish, and brownish sand, in layers dipping southward	2ft to 10ft	0in.
(In the lower portions small ferruginous crusts and masses of conglomerate in a sand or clayey matrix are sometimes found in various pits)		
Stiff blue clay	1ft to 1ft	6in.
Milk-white sand occasionally clayey in the upper part		
Bed of pebbles in which stream-tin is said occasionally to occur		

The following section on the North of the Beacon is given by Dr. Boase (Trans. R. Geol. Soc. Corn. vol. iv. p. 296):

(9)	1	Subsoil—earth with angular stones ...	1ft. to 2ft.	0in.
	2	Yellow and white sand, with minute particles of schorl		
	3	Dark ochreous-coloured sand, with a minute quantity of clay between the grains ...	2ft.	0in.
	4	Soft and greasy, tough, adhesive blue clay, with an oily rancid smell, as if from impregnation of animal matter ...	1ft.	0in.
	5	Clay (called Furnace clay), white and plastic, emitting an argillaceous odour ...	3ft.	0in.
	6	Sand nearly pure white ...	7ft.	0in.
	7	Loose rubbly layer, like (1), said to rest on solid rock.		

(10) Quoted by Mr Henwood (*op. cit.*) from the Mining Review, paper by Mr. Thomas:—Section on the North of the Beacon, half a mile from it, near Wheal Kind. Surface 383 feet above high water. Eight feet sunk in white sand (7 feet below the surface).

(11) Mr. Henwood quotes (*op. cit.*) the following:—N.W. from the Beacon. Surface at 377 feet above high water. Sand met with at 3 feet below the surface: 15 feet sunk through yellow sand.

(12) Messrs. Kitto and Davies give the following section to N.W. of the Beacon:

Soil and Head ...	4ft.	0in.
Blue fire clay (coarse, through admixture of sand) ...	7ft.	0in.
Candle clay, adhesive and very tough ...	2ft.	0in.
Sand resting on Killas ...	5ft.	0in.

(13) Mr. Hawkins (*op. cit.*) gives the following section on the East of the Beacon:

Depth of the deposit, 24 feet in all

Yellow Cobb under vegetable mould ...	2ft. to 3ft.	0in.
Yellow sand ...	3ft. to 4ft.	0in.
Mining clay ...	1ft.	2in.
White sand ...	4½ft. to 5ft.	0in.
A few flattish pebbles in black mud (local name) ...	2ft. to 3ft.	0in.

(14) Messrs. Kitto and Davies give the following section on the east of the Beacon (*op. cit.*):

Head ...	6ft.	0in.
White candle clay ...	3ft.	0in.
Gravel ...	1ft.	0in.
White candle clay ...	2ft.	6in.
Yellow and whitish sand not gone through		

The following sections were taken in the isolated part of the deposit on granite to the West of the Beacon.

(15) Hawkins (*op. cit.*)—

Yellow Cobb ...	4ft.	0in.
Clay ...	6ft.	0in.
Puddle sand (local name) ...	2ft. to 3ft.	0in.

(16) Henwood, quoted from Mining Review (*op. cit.*):—Surface 418 feet above high water. Sand met with at 9 feet below the surface; 12 feet sunk through yellow sand of a lighter tint at the base.

(17) Messrs. Kitto and Davies (*op. cit.*):

Head	9ft	0in.
Candle clay	20ft.	0in.
Dark red sand	0ft.	3in.
Yellowish sand	2ft.	0in.
Gravel, pebbles, boulders, and sand resting on granite	3ft.	0in.
	34ft.	3in.

I observed four pits, all in the main deposit, and lying to the northward of the Beacon, varying from seven to eleven feet in depth; the very impersistent nature of the clay and of the colours in the sands was very noticeable.

From the map and sections accompanying the paper by Messrs. Kitto and Davies (*op. cit.*), it will be seen—that the clays are in no place coextensive with the sands, although in parts their boundary approaches very near to the limits of the deposit; that they are the thickest in the isolated patch on the granite (17), which lies in a basin; that the coarse detritus is of exceptionally local character in the different sections, tin stone pebbles being confined to the immediate vicinity of lodes. The appearances of bedding in the sand, and the relative positions of the sands and clays in sections (1), (7), and (8), are indications of a continuity of deposit, which the variability of the other sections given shows to be abnormal. The occurrence of quartz pebbles in exceptionally sandy overburden (section 6), is worthy of note, and suggests the former overspread of gravelly detritus, similar to the gravels of Crousa Down and Crowan.

The Head, as far as I observed it, consists of brown loam, with angular fragments of local rocks derived from the hill above, resembling, according to Messrs. Kitto and Davies. "The soil and subsoil found upon the Killas of Cornwall, except that it is somewhat sandy in parts and occasionally contains washed pebbles." This Head, Overburden, or Cobb is of like nature, and probably roughly contemporaneous with the accumulations of stony loam on the coasts hereafter to be noticed. The preservation of the deposits in their present form is probably largely due to this protecting envelope of talus shed from the adjacent Beacon hill, which exceeds 600 feet in height.

From such local materials as the granite on the west of the Beacon, the Elvan Course to the north of it, and the Killas, the sands and clays seem to have been formed.

The position of the deposits with reference to the present coasts, and to the high land of the Beacon, and the cliff-like sections and waterworn hollows noticed in some parts, would seem, "as De la Beche suggests," to justify a marine origin, but with them "the resemblance to the raised beaches appears to terminate" (Report, page 258.) The interchangeable characters of the sands and clays are more in accordance with the irregular deposition of a stream, subject to fluctuations attendant on meteorological changes, than with the more uniform sorting action of a coast-fringing sea. The very local character of the basement gravels is also against the admission of a marine origin. As an entirely new system of drainage has been moulded since the deposits were thrown down, proximity to the present coast-line is no argument in favour of marine origin or former proximity to the sea.

Fluviatile agencies, which have produced similar effects in wearing the surface of the shelf in stream-tin sections, coupled with the weathering and water-wear of a vertical face, slickenside, or joint, might, in the absence of further evidence, explain the phenomena of the smoothed surfaces, water-worn hollows, and old cliff face mentioned by Messrs. Kitto and Davies. Had such action prevailed for a long period in an old line of drainage down which the coarser detritus had been swept, the damming up of the old stream course and selection of a new one above the present site of the deposits, would tend to the formation of a lake in whose quiet waters the finer debris of the adjacent land borne down by rills and streamlets would have been filtered, and have settled down in the form of sand and clay.

The isolated positions of the deposits of Crousa Down, Crowan, and St. Agnes, afford no clue as to their relative ages. Yet this isolation justifies me in classifying them together as the oldest superficial deposits as yet noticed in Cornwall. An entire bouleversement of the levels of their respective districts has taken place since their formation, and all traces of synchronous deposition have been swept away in the elaboration of the present drainage system. As they can only be regarded as relics of much more widespread deposits, the possibility presents itself that we may have in them the traces, *in situ*, or re-distributed, of Tertiary or even late Cretaceous deposits, presenting a different aspect to that in other areas through dependence on local sources of supply. During the vast period that intervened between the Culm-measure rocks and the Pleistocene Age, it is unreasonable to argue from the absence of deposits of intermediate age that Cornwall was never invaded by Secondary or Tertiary seas.

On the Occurrence of Flints in Cornwall.—De la Beche (Report, p. 429) commented on the abundance of rolled Chalk flints in the recent as well as the Raised beaches on the Cornish coast; he suggested the existence of a race making use of flint implements prior to the raising of the beaches, and that these flints in transport from the localities whence they were derived, might have been dropped, and, in unlading, have been lost and rolled with the beach pebbles. This theory may be dismissed as untenable both on account of the absence, in inland localities, of relics of such a race as that invoked, and on account of the number of natural flints and the absence of signs of manufacture.

Mr. Peach notices (T.R.G.S. Corn. vol. v. p. 55) the abundance of flints in some of the coves at Gorran, and suggests their derivation from the Chalk of "No Rest," off the Dodman Point, "a name given to some submarine rocks by the fishermen, owing to their trawls becoming hitched in the rough ground."

It is scarcely credible that such observers as De la Beche, Borlase,

Boase, Carne, Henwood, etc., could have failed to notice the existence of Cretaceous rocks off the Cornish coast, and, if known to them, they would certainly have commented upon them. Therefore, in the absence of further particulars, it is safer to regard the "Chalk of No Rest" as a local epithet without any geological significance.

De la Beche, quoting Borlase (Nat. Hist. p. 106, in Report, p. 646), says: "In the low lands of the parish of Ludgvan, in a place called Vorlas, there is a bed of clay, about three feet under the grass, in which numbers of chalk flints are found, with pebbles of quartz and some shingle, with pieces of angular slate." I was unable to find the locality indicated, the present rector of Ludgvan being ignorant of the name. Thinking, however, that Vorlas might be a misprint for Crowlas, a small village on the flats near Ludgvan, I made inquiries there, but failed to elicit any information respecting the occurrence of flints in the neighbourhood.

Mr. Henwood (Journ. R. Inst. Corn. vol. iv. p. 214) mentioned the occurrence of flints of considerable size in the tin ground at Lower Creamy, a part of Red Moor, in Lanlivery, N. of St. Austell. He also stated that a few flints have been very rarely found in a peat bed, containing remains of furze, alder, oak, and hazel, in the stream works of Pendelow, as shown in 1873 (*op. cit.* p. 213).

Mr. Higgs (T. R. G. S. Corn. vol. vii. p. 449) gives a short notice of the discovery of a substance resembling a chalk flint in a cavity in a lode in Balleswhidden Mine.

If the above are Cretaceous flints, and not fragments of slate or fine grit, to which contact with igneous matter had imparted a cherty character, they would seem to indicate the destruction of Cretaceous material, or of deposits of a later date, resulting in part from the waste of Chalk.

Mr. A. Smith (T. R. G. S. Corn. vol. vii. p. 343) mentioned the occurrence of comparatively unworn chalk flints, and fragments of Greensand rock more worn, on Castle Down, in Tresco, one of the Scilly Group.

Mr. Spence Bate (Trans. Dev. Assoc. for 1866) alludes to the occurrence of flints in moorland around Dosmare Pool, Cruza (? Crousa) Down, on the top of Maen rock, at Constantine, and on Trevose Head.

The flints occurring in the Raised Beaches will be noticed in the section devoted to the latter further on.

As the present drift of shingle from W. to E. is the reverse of that which the presence of chalk flints in the recent beaches would lead us to expect, we may conclude that they were obtained by the destruction of the raised beaches, and explain their occurrence in the latter by either of the following hypotheses: first, that the set of the wind-waves during the formation of the Raised Beaches was the reverse of the present, as Mr. Godwin-Austen suggests (Q. J. G. S. vol. vi. p. 87), or, secondly, that during the Pliocene or part of the Pleistocene Period, prior to the formation of the raised beaches, the land stood at a much greater elevation, and the English Channel valley as dry land "served to connect the British Islands with France, etc." (Godwin-Austen, *op. cit.*), that a large part of its area was drained by rivers and streams flowing westward, and carrying Cretaceous and other easterly derived detritus in that direction which detritus, on the submergence of the valley, was incorporated by the Pleistocene sea in the beaches then successively marking its advance, till the culmination of the subsidence at levels marked by the Raised Beaches

Notes on Glacial Hypotheses.

Although the Glacial epoch has left no direct evidences of its changes in Devon and Cornwall it is scarcely possible that either county remained uninfluenced by them The very fragmentary relics of deposits formed during the existence of a previous and very different configuration seems to call for some such powerful denuding agencies as torrential surface waters, consequent on the termination of rigorous conditions of climate.

The Rev. O. Fisher (GEOL. MAG. 1873, Vol. X. p. 163) ascribes the reversal of laminæ in schorlaceous granite, in Carclaze Mine, to the passage of ice over them But such phenomena, as I have elsewhere (Q.J.G.S. 1878, vol. xxxiv. p. 49) endeavoured to show, furnish no proofs of ice-action in the South-west of England. Striæ or moutonnéed surfaces have not been detected in Devon or Cornwall. The grooved face of rock near Barlynch Abbey, North Devon ascribed by Prof. Jukes to ice-action (GEOL. MAG. Vol. IV. p. 41 *vide* Whitley, 32nd Ann. Rep. R. Inst. Corn.), is merely a voluted bedding plane, a structure not unfrequently met with in Devonian and Culm-measure rocks and exhibited by some beds in an adjacent quarry.

If Cornwall was at any time subject to extreme glacial conditions its highlands were not submerged during the Glacial epoch, nor were its borders invaded by a foreign ice-sheet, for traces of submergence would be found in the one case, and foreign ice-borne materials in the other. Positive evidences of local glaciation are also wanting, unless we regard the presence of large boulders at high levels, as the diallage blocks of Crousa Down for instance, as the unremovable *débris* of an old glacier system, and ascribe the presence of large boulders, at some distance from their parent rocks, in river gravels, to the relics of moraine, carried down to successively lower levels in the excavation or deepening of the present lines of drainage. However if, as I agree with Mr. Godwin-Austen in thinking (*op. cit.*), the land stood at a much greater elevation during the Glacial epoch, a great and constant snowfall may have given rise to local glacier systems, and as the present area of the county would offer little more than the generative sources of the (imaginary) glaciers, all traces of pre-existent deposits and of moraine matter, except very large boulders, would be swept down by the flood waters of the succeeding period of subsidence to levels now submerged. But, as all such glacial theories are purely hypothetical, it behoves one to fall back on the probability that Cornwall, during the Glacial epoch, stood at a much greater elevation, and that its highlands were crowned with constant snows, the melting of which during the succeeding amelioration, accompanied by subsidence, caused the liberation of great quantities of surface water with torrential power carrying off the pre-existing detritus to lower lands, now submerged.

PART III.

THE RAISED BEACHES

AND ASSOCIATED DEPOSITS OF THE CORNISH COAST.

THE following observations of the Cornish Cliffs are given in order, proceeding round the coast from Plymouth. The numbers and letters have been prefixed to facilitate subsequent reference.

1. Mount Edgecombe, near Plymouth.

a. De la Beche (Geological Manual, p. 159) mentions the occurrence of rolled shingles, covered by fragments of slate and red sandstone near Redding Point; the height of the deposit is not given.

b. Near Mount Edgecumbe Obelisk I noticed brown and reddish coarse-grained sand filling an inequality in the limestone at about 30 feet above the river; this is probably a trace of contemporaneous deposition with the Hoe Raised Beach.

2. Looe Island. Mr. Pengelly (Trans. R. G. S. Corn. vol. vii. p. 118) noticed the occurrence of layers of comminuted, and somewhat rounded, yellowish matter containing rather large rounded slate fragments and ordinary pebbles, on the northern cliffs of the island. Height above high water not given.

3. St. Austell's Bay.

a. A point at which Raised Beach is engraved on the map, at Polkerris, is capped by 8 feet of Head of small angular killas fragments; occasional quartz pebbles were found, being either the relics of a raised beach, or hurled to a height of 30 feet above high-water mark by storm waves from the beach below. This point is joined to the main cliff by a very narrow ridge of rock.

b. Near Polmero the Head rests upon micaceous slates, and in places presents a rudely stratified appearance.

c. Near the Par Inn, a stratified gravel of subangular grit, quartz, slate, and granite stones, and occasional boulders. 4 to 5 feet in thickness, occurs at about 20 feet above high water.

d. On the south side of Spit Point, fine gravel with pebbles of quartz and boulders (one flint pebble found, and a fragment of *Cardium*, ? *in situ*) 8 feet in thickness, and at base 5 feet above high water, occurs on the low cliffs.

e. Near the above the base of the raised beach is 10 feet above high water; it consists of fine gravel alternating with greyish sand, upon large pebbles and unworn blocks of the subjacent rock. The deposit is 10 feet in thickness, the layers appear to dip seaward.

4. Gerran's Bay.

a. On the eastward side of the beach the section consists of—

Brown soil with angular stones	5ft.	0in.
Brown loam with angular fragments of slate and quartz	10ft.	0in.
Beds of consolidated black sand and quartz gravel, lying unevenly on the subjacent rock at about five feet above high water	4ft.	6in.

De la Beche (Report, p. 430) mentions the consolidation of portions of the raised beach in Gerran's Bay by oxide of iron. Near Pendowa the beach is absent, and the Head rests directly on the slates.

b. Mr. Trist (T. R. G. S. Corn. vol. i. p. 111) described the raised beach as a flat stratum of sand and pebbles, sometimes occurring as a black sandstone 2 feet in thickness, sometimes as a conglomerate of sand and pebbles 10 feet thick, resting on limestones and argillaceous schists abounding in manganese, and capped by an argillaceous friable earth.

c. Near Pendover (? Pendowa) beach, Mr. Trist noticed quartz boulders at the Carnes wholly insulated, and of a different nature from the substratum (*vide* T. R. G. S. Corn. vol. vi. p. 91. Budge.)

d. Dr. Boase (T. R. G. S. Corn. vol. iv. pp. 270, 273) mentions the occurrence of "layers of different substances" in the cliffs to the east of Porthscatho and in Gerran's Bay, the inferior 10 feet being much consolidated. One ferruginous layer resembled pudding-stone. The pebbles diminish upwards into pure sand, reddish brown and friable, in layers 8 or 9 inches thick.

e. (*op. cit.* p. 275.) At Porth, one mile east of St. Anthony, Dr. Boase noticed beds of sand and gravel; Porth farmhouses being built on diluvium of regular beds of sand and pebbles, the latter below; shells, chiefly marine univalves, were found in parallel layers in the sand. The height above high water is not given.

5. Falmouth.

a. Coast-section on the N.E. of Pendennis Castle. Head of angular fragments of slate and quartz with a tolerably regular horizontal lie, 40 to 50 feet in thickness, contains here and there a few pebbles at its base, which is from 5 to 10 feet above high water. Mr. Godwin-Austen mentioned (Q. J. G. S. vol. vii. p. 121) the occurrence of 30 feet of Head on the west of Pendennis Point.

b. Near Cove Battery the Head is of a greyish colour in the upper part, brownish below; a line of larger fragments and a band of loam without stones occur in it.

c. Mr. R. W. Fox (Phil. Mag. and Journ. Science, ser. 3, vol. i. for 1832, p. 471) describes the Falmouth raised beach as—a horizontal bed of rolled quartz pebbles, gravel and sand (like the present beach), from 1 to 3 feet in thickness, and generally from 9 to 12 feet above the highest spring tides. The Head upon the old beach is described as earth, stones, and detached pieces of rock. The cliffs are from 30 to 60 feet in height. The old beach does not extend far from the cliff face; it was observed in one place at 8, in another at 20 feet, within it. Between the parishes of Budock and Mawnan the pebbles appeared to be cemented into a conglomerate, in places, by the oxides of iron and manganese.

d. Mr. Godwin-Austen (T. G. S. ser. 2, vol. vi.) describes the old beach and overlying Head at Swanpool as—purely marine beds passing up into fluvio-marine and fluviatile accumulations.

e. Between Pennance Point and Maen Porth (Fig. 1), a bed of pebbles, chiefly quartz, with slate boulders, is visible, under Head of angular fragments in loam, at intervals. In one place the beach consists of quartz pebbles in grey and reddish brown sand, with large worn blocks of slaty rocks; it is 3ft. 6in. thick, and about 4 feet above high water at its base. Rock platforms are noticeable at about the level of spring tide high water.

Fig. 1.—The Coast toward Rosemullion Head; showing Rock Platforms and Cliffs composed of Head upon Raised Beach.

6. South of the R. Helford.

a. At Ligwrath, between Nare Point and Porthalla, the Head consists of brown earth with angular stones; pebbles are met with in places at its base, at about 5 feet above high water. Boulders compose the present beach.

b. South of the above, traces of a raised beach consisting of beds of coarse black and brown sand, with grit, slate, igneous rock, and small quartz pebbles, in places 2 to 3 feet thick, and at base about 8 feet above high water, are visible here and there under Head of grey and brown loam with angular stones.

c. De la Beche (Report, p. 431) figures part of a consolidated raised beach forming the root of a cavern in the slates on which it rests, and supporting a Head of angular fragments, between Porthalla and the Nare Point. He also gives a sketch of the old beach at Nelly's Cove and between Rosemullion Head and Mainporth (*op cit* p 432)

d. The Rev E Budge (T. R G S Corn. vol. vi p 1) mentions the occurrence of a raised beach, about 5 feet above high water, continuing for some hundreds of yards from Nelly's Cove (¼ mile from Porthalla), and accessible only at low water, he observed traces of the old beach on steep rock ledges now overflowed by the tide. On the north of Nare Point, 8 to 10 feet of angular debris rested on the old beach

7. Coverack Cove

a. The low cliffs to the east of Carnsullan are about 15 feet in height, and composed of brown earth with angular and subangular stones and boulders.

b. The Rev E Budge (*op cit*) describes the cliff-section on the north side of the Cove as—Reddish-coloured marl or rubble upon a thick bed (12 feet) of fine ferruginous sand, consolidated in places, upon large rolled pebbles arranged in regular lines and about 5 feet above high water at their base

c. The same observer says that the whole of the outer portion of the Lowlands in St Keverne parish (a flattish tract of 60 acres in extent) is formed of very fine sand (valued for constructing moulds for brass casting), so similar to that overlying the Coverack raised beach that he considered them contemporaneous. At and near the coast-line pebbles were occasionally met with in the sand.

d. Mr Budge mentions a rampart of large diallage pebbles round a low fortress of sand upon the present beach at Coverack.

e. Dr. Boase (T. R. G. S. Corn. vol iv p. 329) mentions the occurrence of diluvium of an ochreous colour, consolidated toward its base, and containing small pebbles of quartz, compact felspar, and serpentine, resting on serpentine, near Coverack Quay

f. De la Beche (Report, p. 429) and Godwin-Austen (Q. J. G. S. vol. vii p. 121), comment on flints occurring in the Coverack raised beach. Flints also occur in the present beach at Porthbeer Cove, south of Coverack

8. Gunwalloe. The cliffs are capped in places by a Head of light brown loam with angular stones

The Lizard District south of a line between Porthbeer Cove and Mullion was not observed by me, nor can I find any descriptions of Pleistocene phenomena on its sea-board

The low cliffs to the south of the Loo bar are capped by about 5 feet of brown loam with angular fragments of quartz, etc., under coarse brownish blown sand

9. Coast from Loo Pool to Marazion

a. De la Beche (Report, p. 430) figures part of a raised beach between the Loo Pool and Cove village, stained by black oxide of iron, and containing strings of the same substance, the prevalence of which in the rocks of South Cornwall is pointed out.

b. Mr. Henwood (T. R. G. S. Corn. vol. v. p. 54) noticed patches of granite and slate pebbles, from the size of a nut to a foot in diameter, in Tremearne Cliff. The deposits rested on slates at 14 feet above the present beach, in one spot, and at 30 feet in another, going eastward.

c. (*op. cit.*) "At Wheal Trewavas, where the rock is wholly composed of granite, it is covered by a thick bed of transported fragments of micaceous slate."

d. On the west of Pra Sands, Mr. Henwood (*op. cit.*) noticed a bed of granite, elvan, and slate pebbles, at about 6 feet above the present beach, and covered by "a high bank of rubbish," the debris of the adjacent rocks.

e. Between Cuddan Point and Trevean Cove, the Head consists of dark grey loam with angular (local) fragments.

f. The Perran Sands are bounded by cliffs, from 5 to 20 feet high, partly composed of brown loam with angular stones and blocks of greenstone.

g. In a cove west of Perran Sands and south of Perranuthno; in one part—

Brown earth with large and small angular stones	10ft. to 15ft.
upon—large pebbles and subangular fragments of quartz and greenstone	1ft.
upon—brown loam with small angular quartz stones and large angular greenstone boulders.	

g'. In another place—

Soil	2ft. to 3ft.
Brown loam with angular greenstone fragments	6ft. to 7ft.
As above, fragments fewer, and, as a rule, smaller	10ft. to 15ft.
Pebbles, and occasionally subangular fragments, of quartz and greenstone	2ft. (about).

resting unevenly upon greenstone, at from 8 to 12 feet above high water.

h. Toward Marazion the cliffs average 20 feet in height, and are composed of a Head of angular slate, quartz, and greenstone fragments in brown loam.

10. South of Penzance.

a. Mr. Carne (T. R. G. S. Corn. vol. iii. p. 229) observed layers of pebbles and boulders from 3 to 6 feet thick, and 40 feet in length, at the junction of the slate and granite at Mousehole. Mr. Henwood gives the height of the above as a little above high-water mark. (*Ibid.* vol. v. p. 110.)

The following are from Mr. Carne's paper (*op. cit.*).

b. At Carn Silver, boulders and pebbles were found in the end of a cavern, 8 feet wide and 12 feet high, once probably filled with them.

c. In St. Loy Cove, under 30 feet of Head of granitic stones in clay, pebbles and boulders were observed, 4 to 8 feet in thickness, 150 feet in length, and at their base at high-water mark. (Present beach composed of granite boulders.—W.U.)

d Boulders were also observed at Polwarnon (? Polguarnon) Cove, Lean Scath, Peduvounder Cove (near the Logan rock), and at the Land's End Hole; but their height above the sea is not given.

e Near Penberth on the east, I noticed a small patch of Head composed of brown loam with angular stones and angular and subangular boulders.

11 Land's End.

a. In Whitesand Bay, near Carn Aire, the Head consists of angular and subangular fragments and boulders of granite in coarse light buff-brown granitic debris (growan), becoming browner and more loamy near the base. The present beach is composed of granite boulders.

b Between Creagle and Aire Points, Mr Carne (*op cit.*) observed 6 feet of boulders and pebbles under 30 feet of clay with granitic fragments. Base of boulder bed at about spring tide high water.

c. On the south of the Nanjulian River (Carne, *op. cit.*) boulders and pebbles occur at 15 feet above high water.

d On the south of Pol Pry (*op. cit.*), a thin bed of boulders at 20 feet above high water.

e In an iron vein at Huel Oak Point (*op. cit.*) boulders were found at 8 feet above high water.

12 Pornanvon and Porth Just.

a In Pornanvon Cove Mr Carne (*op cit.*) noticed 2 boulder beds (in a matrix of calcareous sand, granitic gravel and clay), separated by a mass of solid granite. The westernmost bed being 4 chains long, 10 feet thick, and overlain by 60 feet of granitic debris, that on the east was found to be 9 chains long, 20 feet in maximum thickness, and surmounted by 20 to 50 feet of granitic debris. The boulders vary in size from that of a hazel nut to 3 feet in diameter, no large slate boulders were noticed. The base of the deposit is about the level of very high spring tides. At Porth Just Mr Carne found boulders at 15 feet above high-water mark.

b Mr Henwood (T. R. G. S. Corn. vol. v. p. 13) mentioned the occurrence of rounded stones of granite, from the size of a nut to 2 or 3 feet in diameter, with a few slate pebbles, and with granitic sand filling the interstices, at from 15 to 20 feet above high water, at Porth Just and Pornanvon. He says that an adit at Wheal Besans Lode, Little Bounds Mine, was driven for several fathoms through one of these beds, which was found to be from 60 to 70 feet in thickness. (In this estimate the overlying Head was probably included.—W. U.)

c Miss Carne (T. R. G. S. Corn. vol. vii. p. 371) stated that the adit of a mine south of Kennal Point enters the cliffs under a mass of pebbles and boulders.

13 Cape Cornwall.

a. In the south part of Priest Cove I noticed a few pebbles and subangular stones (one of granite), in olive-brown loam, and, occasionally, greyish sand, under 50 to 60 feet of Head which presents a stratiform appearance through unequal distribution of fragments, and different tints.

b. In a little cove just north of Cape Cornwall, I observed the following section (Fig. 2) :—

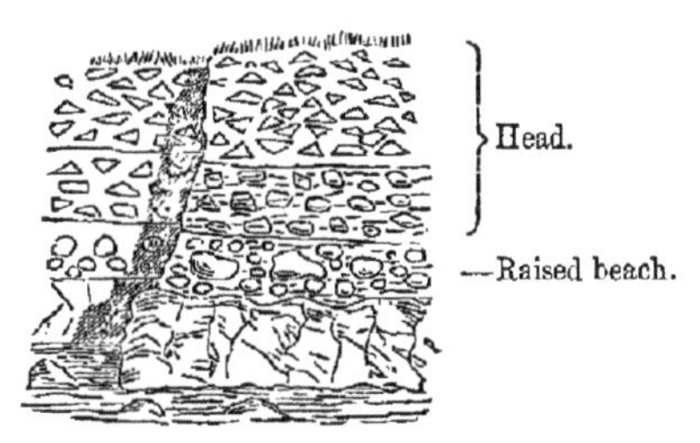

Fig. 2.—Cape Cornwall on the North side.
1 Inch=24 Feet.

Head, brown loam with numerous angular stones, containing larger fragments in the lower 5 feet, with pebbles here and there at and near the base	13ft. 0in.
upon—gravel of pebbles and subangular fragments of slate (altered), quartz, greenstone, a few of flint, and rounded and subangular granite boulders, in coarse brown and black loamy sand	5ft. 0in.

Base of the deposit about 6 feet above high water. Boulders on the present beach. Rock platforms are visible at about high-water mark.

c. In Porthleden Cove the following section was taken :—

Head, brown loam with small angular pieces of quartz, containing small fragments of slate, and, occasionally, granite, 12 feet thick; upon yellowish-brown and brown loam with a few angular fragments; upon well-worn and subangular boulders with a few large pebbles, a few feet above high water.

d. Mr. Godwin-Austen (Q. J. G. S. vol. vii. p. 121) notices the occurrence of granite pebbles, under yellowish clay, with large and small angular stones, and from 5 to 20 feet in thickness, at Creek Tor, in the parish of St. Just, Penrith.

e. On the north of Cape Cornwall, Mr. Carne (T. R. G. S. Corn. vol. iii. p. 229) noticed a bed of slate boulders, 2 feet thick, and a chain in length, on greenstone at 10 feet above high water. The boulders were imbedded in clay and sand with small slate particles.

14. Pendeen Cove (*op. cit.*). Mr. Carne observed 3 feet of small pebbles in sand, made up of comminuted marine shells and pulverized granite, in one place capped by a bed of sand, overlain by 60 feet of Head. The base of the deposit is at about the level of spring-tide high water. The sand is in process of consolidation by iron oxide; it appears to have been blown from the beach into the interstices of the gravel.

15. St. Ives.

a. On the east of Carrack Olu Point, a bed of pebbles, 1 foot thick, is shown under Head, at from 2 to 5 feet above high water. The greenstone composing the Point is capped by a Head of yellowish-brown loam with angular fragments of greenstone.

b. In the bay east of the above, near the north part of St. Ives, the section is as follows :—

Head, with large angular fragments	5ft.	0in.
Impersistent strip of yellowish-brown loam		
Head, loam with a few subangular fragments, and boulders toward the base	4ft.	0in.
Olive and yellowish sand with occasional pebbles	10ft.	0in.

At base about 5 feet above high water; resting upon dark bluish slaty grit with numerous joints.

c. On the north part of St. Ives Island, the greenstone is capped by a Head of angular greenstone fragments from 10 to 15 feet in thickness.

d. Mr. Whitley (Journ. R Inst. Corn No. 11, p. 184) gives the following section of the raised beach in Porthgwidden Cove, St Ives :

Greenstone soil, upon Head of large angular blocks of hornblendic rock Fine sand and loam, upon pebbles of hornblendic rock, quartz, granite, and a few worn flints, mixed with sand, and containing layers of fine brown sand	About 20 feet thick

The base of the deposit is given as 5 feet above high water.

16. Gwythian and Godrevy.

a. Near the southern end of Black Cliffs the slates are capped by a Head of brown clay with angular stones, and a few quartz pebbles at its base.

b South of Ceres Rock, greenish grey slates are capped by a Head of greenish grey clay, probably resulting from their decomposition

c. West of Gwythian ; cliff-section—

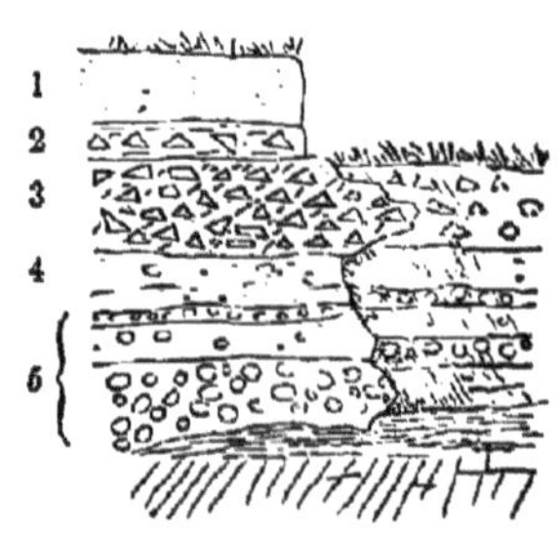

Fig. 3.—Near Gwythian. Vertical scale 1 inch = 12 feet.

1	Blown sand	2ft	0in
2.	Brownish loam with angular slate fragments	1ft.	0in.
3.	Agglomerate of angular slate and quartz stones in a consolidated matrix of small angular pieces of slate	3ft.	0in
4.	Fine brownish sand, consolidated in places, containing a few pebbles	2ft	0in.
5	Three beds of pebbles and subangular stones of slate and quartz, with occasional pieces of flint in the lower bed The beds are 4in, 1ft, and 2ft. in thickness, respectively	3ft.	4in.

d. Near the above, the Head consists of grey loam with angular slate stones of small and average size. The pebble deposits occur in two layers, separated by a seam of brown sand. The base of the gravel is about 5ft. above high water.

The following observations of the Cliffs of Godrevy commence at a point about three-quarters of a mile to the south of Godrevy Island.

e. The section, partially obscured by sandy debris, consists of—

Head, yellowish and grey loam with small angular stones, and occasional large angular quartz fragments, resting unevenly	10ft. to 20ft
upon—fine olive brown sand	5ft.
Coarse grey sand with pebbles and subangular fragments of slate and quartz, the former sometimes large	5ft.
Consolidated coarse blackish sand with small pebbles and subangular fragments, and a few large pebbles...	variable.

At base 5 to 8 feet above high water.

f. The pebble band is stained blackish; it is from 6 inches to 1 ft. thick, and about 6 feet above high water. At this point angular and subangular fragments, some large, are associated with the pebbles in a coarse impure sand matrix.

g. Two beds of coarse blackish and reddish-brown consolidated sand, containing pebbles, etc., of slate and quartz, 3 feet in maximum thickness, and 6 feet above high water at their base, are capped by angular Head. The upper bed forms the roof of a cavern in the slates. (Fig. 4.)

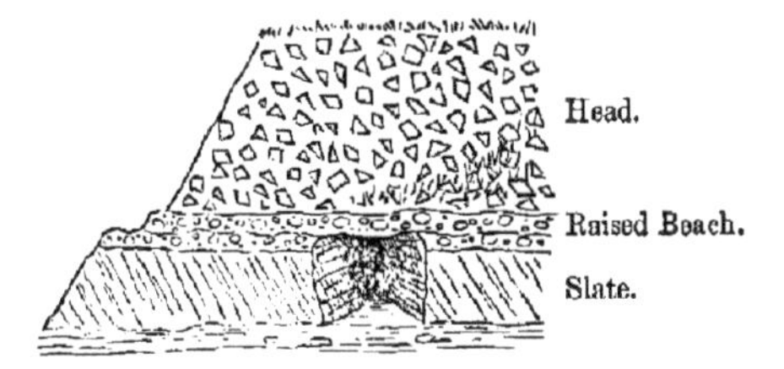

Fig. 4.—Godrevy. Vertical scale—1 inch = 24 feet.

h. A portion of the consolidated raised beach is visible on the foreshore resting upon two bosses of a waterworn slate reef. The denudation of the reef has scarcely affected the unsupported part of the under surface of the beach. (Fig. 5.)

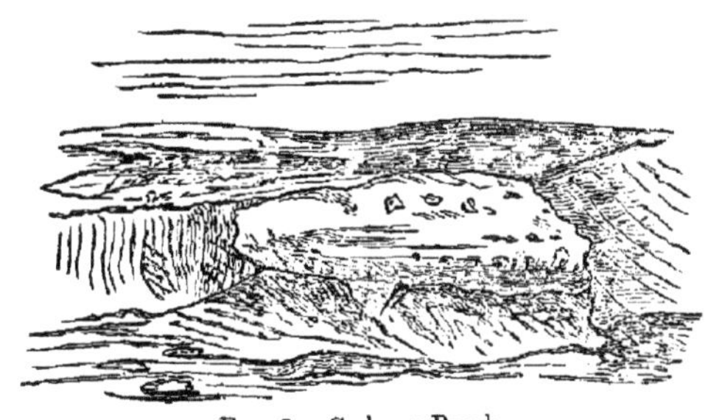

Fig. 5.—Godrevy Beach.
Portion of Raised Beach resting on bosses of Slate isolated from the main cliff.

i. Toward Godrevy Island the beach consists of coarse blackish consolidated sand with pebbles, more gravelly at the base, 4 foot thick, under thick beds of consolidated buff and grey sand with pebbles disseminated through the lower parts.

j. Dr. Paris (T. R. G. S. Corn. vol. i. p. 7) noticed a mass of sand near Godrevy Island, containing whole shells and slate fragments, 12 to 20 feet thick, and 100 feet in length.

k. Dr. Boase (T. R. G. S. Corn. vol. iv. p. 169) described a bed of pebbles above high water, at Godrevy Point and around Fistral Bay, overlain by a bed of testaceous sand; under "transported but unaltered debris," in one place (*op. cit.* p. 309) described as 20 feet of ferruginous clay with angular fragments (local), thinning out landwards as the ground rises.

l. Mr. Whitley (Journ. R. Inst. Corn. No. 11, p. 184) gives a section of the cliffs under Godrevy Farm from top to base.

Brown loam soil	6in. to 18in.
Clay and loam with numerous angular fragments of quartz . .	6ft. to 16ft.
Sandy loam mixed with siliceous sand, and portions of a bed of contorted slate (believed by Mr. Whitley to have been pressed into the bed by ice) .	
Red and white siliceous sand, of quartz grains partially rounded.	
Boulders of blue grit, granite, quartz, vesicular trap (as at St. Minver).	
Slate and a few worn flints in sand cemented by the oxides of iron and manganese.	

m. De la Beche notices (Report, p. 426) the old dunes of consolidated sand, between Gwythian and Godrevy Head, which he distinguished from the underlying raised beach.

17. Observations of the Fistral Bay Cliffs made here and there proceeding northward.

a. South end of the Bay (section obscured in places). Coarse brown semi-consolidated sand, with planes resembling bedding and false bedding, containing occasional lines of small angular slate and quartz fragments, 20 feet thick, seems to underlie Head, shown in a receding part of the cliff. At the base of this old blown sand, a trace of blackish coarse consolidated sand, binding pebbles of slate and quartz, is visible at from 4 to 5 feet above high water.

b. The basement beds consist of gravel of small quartz pebbles, with fair-sized quartz and slate pebbles, and large subangular slate fragments in blackish sand, 1 foot to 18 inches thick, with few pebbles and of a brick-red colour in places, overlain by fine blackish and reddish brown sand with a few pebbles through it, from 2 to 3 feet in thickness.

c. The basement beds are represented by two beds of small quartz and slate pebbles and subangular stones, 6 inches, and from 6 inches to a foot thick, respectively, separated by 18 inches of coarse blackish sand.

d. Coarse consolidated sand of slate and quartz and comminuted shells rests on a pebble bed 2 feet thick, and at base 5 feet above high water. The pebbles are of slate, quartz, and occasionally flint, quartz predominates, the matrix is coarse grey sand.

e. Cliff-section toward the north of the Bay—

Recent blown sand	3ft.	0in.
Sandy soil with angular fragments of slate	2ft.	0in.
Buff loam with angular stones and boulders	2ft.	0in.
Buff sand	1ft.	0in.
Coarse and fine gravel of quartz, dark grey grit, slate and, occasionally, flint	4ft.	0in.

f. Near the above old blown sand is shown, consisting of brown consolidated sand in laminæ about one-eighth of an inch thick, containing pebbles for 4 feet upwards from its base, which is about 10 feet above high water.

g. About 100 yards from the above a trace of consolidated sand binds pebbles at about a foot above high water. Old consolidated blown sand is shown in the cliff above; overlain by Head, capped by recent blown sand. Two whole shells of *Patella* were found near the base of the old blown sand, which forms a tough bedded rock, hardening on exposure to the weather.

h. Dr. Paris (T. R. G. S. Corn. vol. i. p. 7) described the old beaches of Fistral Bay and New Quay as a horizontal bed of pebbles, 10 to 12 feet thick, containing whole shells and slate fragments cemented in sand, resting on slates, and supporting immense heaps of drifted sand.

i. De la Beche (Report, p. 427) describes the Fistral raised beach as rolled pebbles, often large, mixed with smaller gravel and sand, overlain by alternations of fine gravel and sand (the layers being unequally consolidated), capped by sand, becoming mingled with rock fragments, near the extremity of the dunes on the north and south.

j. Mr. Pattison (T. R. G. S. Corn. vol. vii. p. 50) mentions the intersection of the Fistral raised beach by a lead lode, in the middle of the Bay. He describes the present beach as "fine sand and an abundance of shells; it exhibits no pebbles save those derived from the ancient beach."

18. New Quay.

a. On the east side of Towan Head, a trace of black consolidated sand with pebbles is visible at about 6 feet above high water.

b. On the west of New Quay Pier, the section consists of coarse yellowish-brown consolidated sand, chiefly made up of comminuted shells, with a few shells of *Helix;* containing angular, subangular, and rounded stones and boulders of quartz and slate (a granitoid fragment was found) at the base; upon coarser sand with well-rounded fragments resting on a narrow rocky platform 6 feet above high-water mark.

c. Dr. Boase (T. R. G. S. Corn. vol. iv. p. 259) noticed a bed of shelly sandstone, on the north of New Quay signal station, containing fewer shells than at Fistral Bay; the lower part, just above high-water mark, being consolidated into a conglomerate.

d. De la Beche (Report, p. 427) gives a section on the east of Look-Out Hill, New Quay, of an ancient beach of rounded slate pebbles agglutinated by consolidated sand, some feet above the sea-level; capped by layers of sand of comminuted sea-shells consolidated in the lower parts; under a Head of angular fragments from the rocks of the hill above.

19. Between New Quay and Padstow.

a. A thin capping of Head visible on part of Trevelga Head Island.

b. To the west of Tregurrian, Head of angular and subangular slate and quartz stones is shown in the cliffs, under greyish sandy soil.

c. West of Trenance (N. of Mawgan Porth) the Head consists of brown loamy clay with large quartz boulders, and small slate and grit stones.

d. At the north end of Treyarnon Bay the low cliffs are capped by 6 inches of angular quartz and slate stones, under brown clay, one foot thick.

e. The cliffs bounding Constantine Bay, for about three-quarters of a mile, seldom exceed 7 feet in height. Opposite Constantine Island the cliff is composed of—

Blown sand with a layer of broken *Mytili*, and whole *Patellæ*; finer in the lower part, and containing angular pieces of slate, and fragments of shells, as above	4 ft. 6in.
upon—coarse quartzose sand with rounded grains	1ft. to 2ft.

resting on slates at 6 feet above high water.

f. Near the centre of Perleze Bay a few quartz and slate pebbles are present, under blown sand, at about five feet above high water.

g. The cliffs of the cove north of Trevone (2 miles west of Padstow) are from 5 to 15 feet high, and occasionally capped by coarse brownish sand, giving place to dark brown clay with angular slate fragments and an occasional quartz pebble. The ground slopes upward for a quarter of a mile very gradually.

20. Between Padstow and Tintagel.

a. On the cliffs of Bray Hill, near St. Enodock, yellowish and grey, thin bedded, consolidated sand of comminuted shells, containing shells of *Helices*, is visible; at base 5 feet above high water.

b. In one place the following section occurs under 2 feet of recent blown sand:

Consolidated sand	6in.
Dark brown loam, containing angular fragments of quartz, slate and grit	2ft. to 3ft.
Upon greenish grey slates with quartz veins.	

c. Near the mouth of the St. Enodock Valley, a bed of consolidated sand, one foot thick, containing land shells and angular fragments of slate, is capped by recent blown sand, and rests on red and green banded slates at 8 feet above the present beach.

d. At the stream mouth between Porteath and Trefan Head, Head, of angular stones in brown and yellowish loam, has a stratified appearance in the distance, owing to the sizes and dispersion of the fragments and their partial absence in places. The stream has cut a steep bank at the mouth of its gorge, which exposes 20 feet of Head—brown loam with angular slate and quartz stones, roughly horizontal in arrangement.

e. By the mouth of the stream west of Port Isaac, between Roscarrock and Lobber Rock, 20 feet of Head is shown, consisting of angular fragments of slate and quartz in brown loam.

f. Near Chapel Rock the slates have been cut into reef platforms or shelves, in places, at about high-water level.

g. At the mouth of the stream gorge west of Dannon Chapel, 10 feet of Head is shown, consisting of brown loam with angular slate and quartz stones.

21. The Scilly Isles.

Mr Carne (Trans. R. G. S. Corn. vol. vii. p. 140) mentions the occurrence of redistributed granitic matter, called "secondary granite," on Rat Island, at Piper's Hole in Tresco, and Piper's Hole in St. Mary; in both the latter caverns it forms the principal part of the roof, and contains boulders or rounded masses of perfect granite, some rather large.

General Conclusions.

Head.—The position of the stony loam or Head in sites where no modern talus could rest, the denudation it has undergone, and its frequent presence on the cliffs, prove its accumulation to have taken place subsequent to the formation of the raised beaches, yet considerably anterior to the prevalence of the present climatal conditions. It marks, as Mr Godwin-Austen (Q.J.G.S. vol. vii. p. 122) says, "A time when the degradation of the surface proceeded much more rapidly, and when fragments of rock far exceeding the motive power of any rainfall were conveyed down slopes along which only the minutest particles of matter are now carried" (*vide* 9 *c*). Such conditions of long-continued subaerial waste are likely to have prevailed, as Mr Godwin-Austen suggests (Q.J.G.S. vol. vi. p. 93, etc.), during a greater elevation of the (South) West of England.

The rough appearance of stratification sometimes noticeable in the Head [(1) through the horizontal lie and apparent regard to gravity in distribution of its contained fragments, *vide* 3 *b*, 5 *b*; 13 *a*; 20 *d*; (2) through strips of loam or clay without stones as in the higher cliffs bordering Pra Sands, and 15 *b*, (3) through percolation of water carrying down overlying substances to a certain horizon, as 17 *e*; (4) through distribution of colouring matter, as 5 *b*, 13 *a*] may in many cases be due to fluviatile deposition, to which Mr. Godwin-Austen referred the Head at Swanpool (5 *d*) and other places.

We cannot suppose that no fluviatile deposits were formed during this period of subaerial waste, judging from the pell-mell distribution of angular fragments in the torrential gravels of the present streams, in their higher reaches, it is only reasonable to expect that similar deposition would then have taken place on a much larger scale, and that its traces would be found in the present area of the county which would only represent the highlands of its former extension.

Raised Beaches.—The general consolidation of the old beach materials, occasionally into a very hard rock (*vide* 6 *c*, 16 *g*, *h*, 18), renders their detection, even as fragments on a level with the surface of the present beach, comparatively easy; where, however, the process of consolidation was interfered with by the accumulation of the Head the beach material seems to have been swept away, and in some cases to have left traces in occasional pebbles at or near the base of the Head (*vide* 5 *a*; 6 *a*, 16 *a*; and perhaps 19 *g*). Even where the raised beach is well developed, the upper part has been sometimes mingled with the base of the overlying talus (13 *b*). Angular fragments are occasionally found in the raised beaches (16 *f*). The above observations serve to explain the appearance of beach material on Head, S. of Perranuthno (9 *g*), as, in an adjacent section (9 *g*′), the Head is represented by pebbles and subangular fragments. On Bray Hill (20 *b*), 6 inches of consolidated sand rests on 2 to 3 feet of Head; but the latter is represented in an adjacent spot by consolidated sand with angular fragments of slate, and land shells (20 *e*), so that the old sand drift may have taken place on the beach platform after a little talus had been shed upon it during the earliest symptoms of elevation.

It is often difficult, where old consolidated blown sands occur, to distinguish their junction with underlying raised beaches, as pebbles and fragments of *Mytili*, *Patellæ*, etc. (16 *i*; 17 *f*, *g*, 18) may have been cast upon the dunes by storm waves, their presence and linear arrangement in recent blown sand (19 *e*) would seem to be due to protracted gales from the same quarter.

Being chiefly composed of comminuted shells, the percolation of water through the old dunes would best explain their consolidation. Dr. Paris (T.R.G.S. Corn. vol. i. p. 7), in addition to this, gives two other possible modes of consolidation, viz., by water charged with pyritical substances, or by ferruginous infiltration.

The absence of organic remains in the majority of the Cornish raised beaches has been ascribed to Arctic currents (Godwin-Austen, Q.J.G.S. vol. vi. p. 87), which I think very probable. It also suggests the possibility that many of them may have been deposited by rivers, or in estuaries, whose seaward banks have been swept away. Some of the boulder beds mentioned by Messrs. Carne and Henwood are at too great heights to be regarded as raised beaches, and may more reasonably be referred to far older fluviatile deposition. If the adit mentioned by Mr. Henwood (12 *b*) cut through a continuation of the worn boulder beds of Porth Just and Pornanvon to a thickness of 60 feet, the boulder beaches of these localities must be regarded as anterior to the raised beaches. In the formation of the old beach cliffs at the termination of a long period of subsidence, fluviatile deposits (thrown down before, and during, the initiation of the present lines of drainage) would have been truncated, so to speak, and exposed at different heights upon the cliffs, just as we find old river gravels exposed on the secondary cliff-line of Devon.

Again, during the elevation of the old beaches, the existing river channels would have been deepened, and river deposits formed in the breaches of the old cliff-line, to be redistributed by the sea in its recent advance. How far boulder gravels and unfossiliferous raised beaches (provisionally so called) may be referred to either of these periods of fluviatile action it is impossible to say, without a searching investigation of each particular deposit with reference to its surroundings.

The local elevation of the raised beaches cannot be correctly measured by the height above high water of their remains; for such an estimate ignores the original thickness of the beaches, and postulates an identity in the local rise of tide during the raised beach formation and at present. The latter supposition is improbable when we take into account—

1stly. The destruction by the sea, during elevation (of the old beaches), of such inequalities as may have proved obstacles to the stream of tide.

2ndly. The modification the raised sea-bed would have undergone through subaerial agencies.

3rdly. The probably different relations of land and sea in other parts of England, and on neighbouring coasts during the formation of raised beaches in the S.W. counties.

4thly. The subsequent modification of the old coast-line.

Mr. Pengelly (Trans. Dev. Assoc. part v. p. 103) points out the fallacy of supposing—that all contemporary raised beaches are on the same level, and the converse—that raised beaches on the same level are necessarily contemporaneous. The cautions given show the danger of laying stress upon individual observations which may be taken where the beach was left very thin, or at different parts of its seaward slope.

The base of the Cornish raised beaches above high-water is shown by observations to average 5 feet; such cases as Pendeen Cove (14); Tremearne (9 *b*); Nanjulian (11 *c*); Porth Just (12 *a*); being exceptional. Taking the thickness of the old beaches at 15 feet as a maximum, the average subsidence indicated by them would be from 12 to 20 feet below high water.

De la Beche (Geological Manual, p. 157) gives a section of the successive faces (indicated by dotted lines) that the degradation of a cliff composed of Head upon raised beach would be likely to exhibit (see Fig. 6).

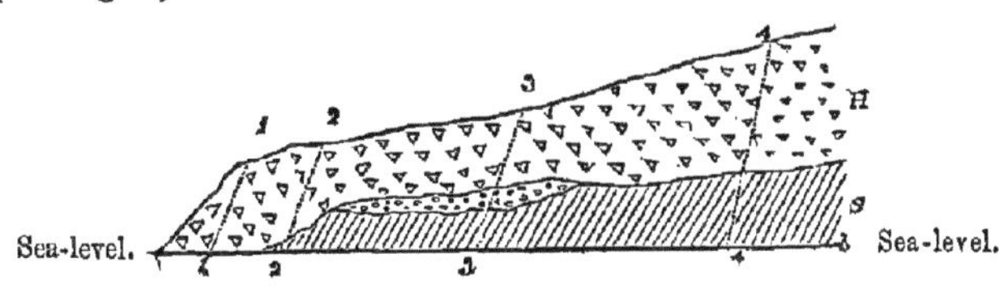

Fig. 6.—H, Head, concealing a raised beach, resting upon slate, S, above the sea-level.

The raised beach platform has been cut too far back to allow of such cliff faces as 1 and 2. Exceptions to this rule may be furnished by the low tract at Spit Point near Par; the lowlands of St. Keverne (7 *c*); the flattish tract covered by blown sand between Constantine and Perleze Bays, if the waterworn sand (in 19 *e*) is a

trace of raised beach, or rests on an old beach platform ; the gentl
sloping tract bordering the coast near Trevone (19 *g*).

The cliffs bordering a part of Pra Sands are wholly compose of Head to a height of 60 feet from the present beach ; but as Hea rests on a portion of raised beach on an adjacent promontory, on platform 5 feet above high water, the old beach platform may i this instance have been broken up by fluviatile agencies prior to o during the accumulation of the Head ; or the original surface of th platform must have been most irregular. Such cliffs as Nos. 3 and are by far the most general sections on the Cornish coast which ha been in very many places cut too far back to show either raised beac or Head.

PART IV.

PLEISTOCENE GEOLOGY OF CORNWALL.

Submerged Forests and Stream Tin Gravels.

THE evidence under this head is necessarily a compilation, th very exceptional exposure of the old forest ground, and th nature of stream tin sections, leaving no room for personal investi gation. The names of the observers are in most cases sufficien vouchers for the accuracy of their statements. The submerge forests are given first, as there is no evidence forthcoming to show the priority of the stream tin gravels to the general growth of th forests. The forest bed overlying the stream tin which Mr. Carn rightly synchronizes with the forest beds on the coast may represen a very brief portion of a long period of forestial growth.

Submerged Forests.—Proceeding round the coasts from Plymouth 1. Looe. Mr. Box (26th Ann. Rep. Royal Inst. Corn. for 1844 noticed trunks of oak, alder, ash, and elm, on Millendreath Beach in vegetable mould extending for 250 yards from east to west, an sloping from below high-water mark to the southward for 150 feet where it was lost sight of under fine sand, which, though explore for 30 feet farther out, yielded no further traces. The plants in th mould resembled those found in a neighbouring marsh, 130 fee above high water, of which the following section is given :—

Peat of flags and arundaceous plants.
Dark brown vegetable matter with holly and alder.
Layer of sand with vegetable matter, numerous hazel nuts, and the elytr of Coleopterous insects, also black oak and ? holly, resting on firm light coloured clay.

Numerous angular slate fragments were met with, but no shells.

2. Near Mevagissey. Sir C. Lemon (T. R. G. S. Corn. vol. vii. p. 29) gives the following section disclosed in cutting a drain at Heligan (about a mile inland from Mevagissey Bay) near the foot of a hill 20 feet above the stream in the valley bottom, and in another place, higher up, at 40 feet above the stream.—Loam 1 foot 8 inches from the surface, upon a mass of whitish, bluish, and yellowish clay with broken slate, with the stump of an oak 4 feet long and nearly a foot in diameter, 7 feet 4 inches from the surface at its lower extremity.

Submerged forests have been observed after severe gales—

3. At Fowey by Mr. Peach (T. R. G. S. Corn. vol. vii. p. 62), the trees being rooted in stiff clay.

4. At Porthmellin, near Mevagissey (*Ibid*, vol. vi. pp. 23 and 51), the roots resting on clay apparently *in situ*.

5. At Maen Porth, near Falmouth, by the Rev. J. Rogers (*Ibid*, vol. iv. p. 481), the roots being in clay.

6. At Porthleven near the Loo Pool, by the Rev. J. Rogers (*Ibid*, vol. i. p. 236), oak and willow roots apparently *in situ*. At Fowey and Porthmellin, elytra of beetles were found.

7. Mr. H. M. Whitley (Journ. R. Inst. Corn. No. 13, p. 77) gives the following section at Market Strand, Falmouth, exposed during excavations at the Landing Pier:—

Layer of sand on a thin bed of shale, thinning out seaward	2ft.	0in.
on—Forest Bed, compact peat, flags, ferns, trees of oak, hazel, fir, beech; fir and beech most abundant; no hazel nuts obtained	7ft.	0in.
The top of this bed occurred at about the level of ordinary spring-tide low-water mark. Its base rested on a layer of gravel	4ft.	0in.

Mr. Whitley was informed that the forest bed extended for a short distance up the valley, and that another part of it had been met with in an excavation at Bar Pools. The open space before the market is called "the Moor."

8 *a*. Mounts Bay. Leland thus alludes to the submerged forest in Mounts Bay—"In the Bay betwyxt the Mont and Pensants be found near the lowe water marke Roots of Trees yn dyvers places as a token of the ground wasted."

b. Dr. Borlase (Trans. Roy. Soc. for 1757, p. 80) noted the discovery of roots, trunks, and branches of oak, hazel, and willow, on the shores of Mounts Bay, in black marsh earth with leaves of *Juncus*, under 10 feet of sand.

c. Dr. Boase (T. R. G. S. Corn. vol. iii. p. 131) mentioned the occurrence of vegetable mould with roots and trunks of indigenous trees, under 2 to 3 feet of sand on the west of St. Michael's Mount.

d. Mr. Carne (T. R. G. S. Corn. vol. vi. p. 230) noticed the occurrence of trees on peat, east of Penzance, the largest being an oak trunk with bark on, 6 feet long and 1½ feet in diameter.

e. He also mentioned the occurrence of a peat bed 3 to 8 feet thick in the low tract between Marazion and Ludgvan (a reclaimed marsh), it extends for 2 miles, from a little eastward of Chyandour to the Marazion River Near Longbridge, where it approaches the surface, it is from 4 to 7 feet thick, and used for fuel, it rests on a thick bed containing *Cardium edule*, and is generally concealed by alluvium

9. Mr. Henwood (40th Annual Rep R. Inst Corn for 1858) describes a submarine forest on Dunbar Sands in the Camel Estuary Nothing save spongy masses of peaty sand were visible in 1875, when I visited the spot, the roots, etc, having been probably washed away in the interim.

10 De la Beche says that traces of submarine forests were noticed at Perran Porth, Lower St Columb Porth, and Mawgan Porth (Report, p 419). No signs of them were visible on the occasion of my visit St. Columb Porth is a sand flat, at low water, between cliffs not 10 feet in height, exhibiting no traces of old marine action. Mawgan Porth is a similar sand flat, but broader, and terminating in low sand dunes, to the south of which narrow strips of alluvium border the streams.

11 Bude Mr S R Pattison (T R G S Corn. vol vii p 35) noticed roots of trees of large size, apparently *in situ*, in dark clay, at Maer Lake, near Bude Haven.

12 Mr Pattison also noticed large accumulations of bog timber in the Fowey Valley on Bodmin Moor At Bolventor the heads of the trees pointed down the valley

Stream Tin Sections.

1 De la Beche (Report, p 405) says that in the interior the tin ground is usually covered by river detritus, more open spaces frequently having a bed of peat (in which oaks are common) interposed between the tin ground and other detrital accumulations, as in Tregoss Moor and the moors adjacent to Hensborough "In some whole ground (stream tinners' term for stanniferous gravel) and superincumbent beds not previously disturbed by the old men, upon Bodmin Const Moor, the peat beds with oak, alder, etc, covering the tin ground very irregularly, were in some places several feet thick, in others absent, though on the whole they seemed to keep a somewhat common level above the tin ground In some places thin peat beds had been accumulated at still higher levels among the gravels, sands, and clays The shelf composed of semi-decomposed granite was very irregular holes 30 or 40 feet deep presenting themselves, in the bottoms of which there was usually good stanniferous gravel"

2 Mr Pattison (*op. cit*) gives a section of the Fowey Valley Works, in which the (hard and black) forest bed was met with at from 23 to 27 feet below the surface, resting on stream tin gravel, and overlain by sand with a peat bed containing ferns and hazel. The granite shelf, tin gravel, and forest bed presented a faulted appearance.

3. Par De la Beche (Report, p 403) In cutting the Par Canal at Pons Mill, near St. Blazey, granite blocks, as if arranged for a bridge, were found beneath 20 feet of gravel, probably in part

resulting from stream tin washing. Section in low ground near the Par Estuary—

1. River deposits ...	1ft.	6in.
2 and 3. Mud, sand, clay, stones, much disturbed by the stream tinners in the upper part, with vegetable matter in the lower part ..	15ft.	0in.
4. Fine sand with sea shells like cockles, and rolled pebbles in the upper part	4ft.	0in.
5. Mud, clay, sand, wood, nuts, etc., mixed ..	3ft.	0in.
6. Tin ground resting upon an uneven surface of slate ...	6in. to 6ft.	0in.

4. North of St. Austell. Mr. Henwood gives the following sections. The letters prefixed denoting beds probably contemporaneous. (T.R.G.S. Corn. vol. iv. pp. 60 to 64.)

A. Merry Meeting, in parish of St. Roche.

a.	Mud, with decayed vegetable matter ...	2ft. to 3ft.
1.	Granitic gravel ...	2ft.
2.	Silt, with decayed vegetable matter and plates of mica ...	4ft. to 5ft.
b.	Granitic stones, gravel and sand mixed with silt and nuts	4ft.
3.	Vegetable matter (locally called Fen), moss, grass, wood (? charred)	1ft.
4.	Silt (vegetable remains ?)	1ft.
5.	Vegetable remains (charred like No. 3) ...	1ft. to 3ft.
6.	Vegetable matter passing into silt ...	8in. to 10in.
c.	Tin ground, with enormous quartz blocks, some 15 ft. square; tin ore as sand, stones, and pebbles mixed with quartz, granite, and schorl rock; little rounded, and of the best quality where the decomposed granite shelf is softest	4ft. to 30ft.

B. In the centre of Pendelow Vale.

a.	Granitic sand and gravel ...	12ft.	0in.
1.	Silt (vegetable matter ?) ...	1ft.	0in.
2.	Granitic sand	4ft.	0in.
3.	Vegetable matter (like No. 5 in other sections, but with sand)	2ft.	0in.
b.	Silt, sand, and gravel mixed	2ft.	0in.
4.	Vegetable matter (like No. 5 in other sections) (Fen)	4ft.	0in.
5.	Tin ground, ore not abundant, most plentiful near the base	5ft.	0in.

C. Watergate.

a.	Mud with granitic sand and gravel ...	5ft.	0in.
1.	Fine granitic sand	2ft. to 3ft.	0in.
2.	Silt (with decayed vegetable matter ?)	2ft.	6in.
3.	Fine granitic sand ...	2in. to 3in.	
4.	Silt (resembling No. 2) ..	3ft.	0in.
b.	Silt, sand, gravel, and large stones, indiscriminately mixed	3ft.	0in.
5.	Vegetable matter passing into silt in the lower part (like Nos. 5 and 6 in the Merry Meeting section)	5ft. to 6ft.	
c.	Tin ground, the ore occurs as sand and pebbles	2ft. to 20ft.	

D. Broadwater Luxillion. Tin ore much larger towards the sea than up the vale. A patch of slate some hundreds of feet in area was found resting on tin ground, and apparently unconnected with the shelf.

a.	Granitic sand ...	6ft. to 7ft.	0in.
b.	Mud, apparently of vegetable origin, mixed with granitic sand and gravel	4ft. to 5ft.	0in.
c.	Tin ground, ore, small pebbles not much rounded	7ft.	0in.

The tin bed is sometimes divided by a bed of granite (cap shelf) as at Grove and Merry Meeting. Numerous blocks of quartz lie on the shelf. Below the shelf (soft granite) tin ore is not abundant.

The following are from Journ. R. Inst. Corn. vol. iv. p. 214:—

E. Levrean in St. Austell's parish.

1. Granitic sand and gravel	1ft	0in
2. Peat (Fen), often mixed with, and sometimes divided by, very thin layers of granitic sand	1ft.	0in.
3. Granitic matter, particles and granules of tin, rarely minute specks of gold (Upper Tin Ground)	3ft. to 6ft.	0in
4. Angular and subangular masses of granite in granitic sand without any tin ore (False Shelf)	1ft to $1\frac{3}{10}$ft	0in
5. Tin ground, angular and subangular granite, felspar, quartz, schorl, veinstone materials mixed with granitic gravel and sand, grains and particles of tin oxide, and less frequently flakes of schistose matter with specks of gold. A few ancient shovels of wood, bound on the edges with iron, have been found in this bed. The shelf is of granite of unequal hardness	10ft. to 15ft	0in

F. Pit Moor in St. Austell's Parish.

1. Vegetable mould	1ft	0in.
2 and 3. Granitic detritus in many layers	5ft. to 6ft.	0in.
4. Tin ground, angular, subangular, and rounded masses of granite, quartz, schorl, veinstones, small quantities of tin ore, clay-slate laminæ, occasional; on soft granite shelf	3ft. to 10ft	0in.

G. Upper Creamy (Wheal Prosper.)

1. Peat	0ft	6in
2. Granitic clay, often mixed with laminæ of yellowish slate	1ft. to 3ft	0in
3. Tin ground, small angular and rounded granitic and veinstone material, tin stone as sand and gravel, microscopic particles of gold. On shelf of bluish and brownish clay. The roots of marsh plants penetrate the tin ground	4ft to 5ft.	0in

H. N.W. of the Railway Bridge over the high road between Lanivet and the Indian Queens.

1. Vegetable mould	6in to 1ft.
2. Angular and subangular stones of quartz, slate, elvan, schorl rock, slate veinstones, and occasionally granite	3in to 4ft.
3. Tin ground like the overburden, but with rounded masses of tin ore, often very small; on shelf of clay slate	1ft to 2ft.

I. Gun-deep in St. Denis.

1. Vegetable mould	6in to 1ft.
2. Gravel, stones of slate, quartz, elvan, schorl rock, and occasionally granite	4ft.
3. Peat	1ft
On 4. Tin ground; poor	

J. On N. side of Tregoss Moor. Ancient works resumed at Golden Stream about half a mile S.E. of Castle-an-dinas in St. Columb Major.

1. Vegetable mould	0ft	6in.
2. Angular and subangular masses of slate, quartz, elvan, schorl rock, veinstones, and occasionally granite; lumps of peat had been previously removed from this bed	5ft to 6ft.	0in.
3. Tin ground resembling the overburden, but with more numerous fragments of elvan, the tin ore as gravel or sand	2ft. to 3ft.	0in

K. Dr. Boase (T.R.G.S. Corn. vol. iv. p. 248) mentioned the occurrence of siliceous sand under diluvial debris in the Stream Works near Hensborough, on the road to Roche. At Tregoss and Roche the tin ground contained quartz and schorl pebbles, and the shelf consists of decomposed slaty felspathic rock.

L. Henwood (J. R. Inst. Corn. vol. iv. p. 230). Section at Penny Snap (Wheal Prosper, in Alternun) E. of the Drains River—

1. Peat	7ft.	0in.
2. Angular and worn granite, elvan, schorl, and quartz stones in pale blue felspathic clay, averaging	5ft.	0in.
3. Tin ground as above, with tin ore as waterworn sand or gravel, on granite shelf	3ft.	0in.

5 A. The section of the Happy Union Works by Mr. Colenso (1829) has been quoted by several writers, but by none more fully than De la Beche, from whom I extract (Report, pp. 401, 402, 403), giving the deposits in reverse order.

1. Rough river sand and gravel, here and there mixed with sea sand and silt. A row of wooden piles with their tops 24 feet from the surface, apparently intended for a bridge, were found on a level with spring-tide low-water ...	20ft.	0in.
2. Sand; trees all through it, chiefly oaks, lying in all directions; animal remains, bones of red deer, hog, human skulls (?), bones of whales	20ft.	0in.
3. Silt or clay and layers of stones, a conglomerate of sand, silt, bones and wood	2ft.	0in.
4. Sand with marine shells, water draining through this bed is salt above, fresh below	0ft.	4in.
5. Sludge, or silt, brownish to a lead colour in places, with recent shells which, particularly the bivalves, are often in layers, double and closed, with the siphonal end upward, rendering it likely that they lived and died there; they are of the same species as those existing in the neighbouring sea; wood, hazel nuts, and occasionally bones and horns of deer and oxen are found in this bed: a piece of oak, shaped as if by man, with a barnacle attached, was found at 2 feet from the top	10ft.	0in.
6. A layer of leaves, hazel nuts, sticks, and moss (in a perfect state, almost retaining its natural colour, apparently where it grew). It extends, with some interruptions, across the valley, occurs at 30 feet below low-water mark, and about 48 feet below spring-tide high-water	6in. to 12in.	
7. Dark silt, apparently mixed with decomposed vegetable matter	1ft.	0in.
8. Roots of trees in their natural position, oaks with fibres traceable for 2 feet deep. "From the manner in which they spread there can be no doubt but that the trees have grown and fallen on the spot where their roots are found." Oyster-shells still remain fastened to some of the larger stones and to the stumps of trees		
9. Tin ground, with rounded pieces of granite, and subangular pieces of slate and greenstone. Most of the tin occurs in the lower part, from the size of the finest sand to pebbles 10lbs. in weight, some rocks richly impregnated with tin weigh 200lbs. and upwards. Thickness (including No. 8) from	3ft. to 10ft.	

B. De la Beche (Report, p. 403) says, "These works are now abandoned," others on S. of London Apprentice Inn were carried on in 1837: "from which it would appear that from the general rise of its bottom, the sea had not entered this valley sufficiently high to permit marine deposits to be there accumulated." This probably refers to Mr. Colenso's section of Wheal Virgin Works (T R G S. Corn vol iv. p 38), a mile from Happy Union, in which no sea sand was found. The tin ground betraying signs of old men's workings lay beneath 32 feet of silt and river gravel, with oak, willow, etc., in considerable quantity, with their roots in situ where soil exists "How far," says Mr. Colenso, "Pentuan Valley extended seawards is conjectural, but at its present declivity of 45 feet to a mile between St. Austell and Pentuan, it must have continued a mile further than it does now." Mr Smith (*Ibid*, p. 400) mentions the rapid descent of the valley from Hensborough (900 to 1000 feet in height), and the continuance of a bed of pebbles all the way.

C. Section of Lower Pentewan work, quarter of a mile from the beach given by Mr Smith (*op cit*)—

1.	Soil with growing trees, some very old, gravelly towards the bottom	3 3
2.	Fine peat, roots of trees, fallen trunks, sticks, ivy, sea laver, rushes, impregnated with salt	12 152
3	Sea mud, with compressed leaves at the top, cockles at 31 feet from the surface, bones, human skulls (one of a child), deer horns At the bottom, a bed of very small shells a foot in thickness	20 35
4	Sea mud, oysters, and cockles	4 39
5	Compressed leaves, vegetable matter, a few rotten shells	6½ 45½
6.	Vegetable matter, rushes, fallen trees, leaves, roots, moss, the wings of Coleopterous insects	1 46½
7	Moss, hazel nuts, sticks, on pebbles of killas, growan, etc	3 49½
8	Rough tin ground, stones light and poor	2 51½
9	Rough tin ground, rich stones with quartz pebbles and yellow ferruginous clay. Killas at about low-water mark	3 54½

D (*op cit*) Section of Upper Pentuan works, 1 mile N from the beach where the valley is half a mile wide

1.	Soil with trees growing on it	3ft.	3in
2	Mud with gravel seams resembling false bedding	21ft	11in.
3 and 4	Spar and killas upon growan, spar, and killas	12ft.	9in
6	Gravel, with trees and branches of oak of great size at the bottom	8ft	0in
6	Tin ground	8ft	5in
7.	Clay, in which were found the roots of a vast oak, and a branch 4 feet long and 3 inches in diameter, projecting from the wall of the work A second mineral deposit may occur below this		

E. Mr Smith also gives a section of Pentowan work (either a place near Pentuan, or a misprint) in 1807.

Sandy clay, stones, gravel	9ft	0in
Peat with roots and leaves	7ft	0in
Sand with branches and trunks of trees	8ft	0in
Finer sand, with shells, bones, horns, vertebra of a whale, human skulls	12ft	0in
Coarse gravel	2ft	0in
Close sand with clay, becoming peaty near the base	12ft	0in
Loose stones and gravel, 1 foot thick, resting on tin ground		

Falmouth district.

F. Tregoney Stream Work in 1807, given by Mr. Smith (*op. cit.*).

1.	Granitic gravel with layers of sand ...	11ft	6in
2.	Black mud with shells (a cow's horn and horns of stags)	15ft	0in.
3.	Tin ground ... ... averaging	2ft.	0in.

6. In Journ. Roy. Inst. Corn. vol. iv. p. 204, etc., Mr. Henwood gives the following sections in two places, where the bed of Restronguet Creek is some 12 feet below spring-tide high-water.

A. Section 1.—

1.	Mud of the river, very soft ...	6ft.	0in.
2.	Mud and coarse sand ...	8ft.	0in.
3.	Mud (hardened) ...	6ft.	0in.
4.	Mud (with numerous oyster shells) ...	12ft.	0in.
5.	Mud (hardened) ...	31ft.	0in.
6.	Tin ground, 6 inches to 6 feet thick ... averaging	4ft.	0in.
	Shelf of buff or blue clay slate		

B. Section 2.—

1.	Soft river mud ...	7ft. to 9ft.	0in.
2.	River sand and mud ...	9ft.	0in.
3.	Blue mud (shells of oyster, cockle, etc.) ...	9ft.	0in.
4.	Stiff blue mud without shells ...	36ft.	0in.
5.	Tin ground, subangular masses of granite, slate, elvan, quartz, etc., and tin ore in large masses interspersed with smaller grains, 6 inches to 6 feet thick; ... averaging	4ft.	0in.
	Shelf of clay-slate.		

De la Beche (Report, p. 403). Up the Carnon Valley in the direction of St. Day, the tin ground is partly covered by marine sediments, partly by common river detritus.

Carnon. Mr. Carne mentioned (T.R.G.S. Corn. vol. iv. p. 105) some beds of slate found reposing on the tin ground in the Carnon Valley, unconnected with the sides and bottom.

C. Mr. Henwood (T.R.G.S. Corn. vol. iv.) gives the following section of Carnon Stream Works, the letters denote beds probably contemporaneous with those in the Watergate, Merry Meeting, and Broadwater sections.

a	Sand and mud; 2 beds; river wash ...	3ft.	0in.
2.	Silt and shells, 3 successive beds ...	0ft.	10in.
3.	Sand and shells (a stream of fresh water percolates through this bed) ...	2ft.	0in.
4.	Silt; 3 beds ...	12ft.	0in.
5.	Sand and shells ...	3ft. to 4ft.	0in.
6.	Silt with numerous shells ...	12ft.	0in.
7.	Silt with stones in places ...	18ft. to 22ft.	0in.
b.	Wood, moss, leaves, nuts; dark coloured as if charred, a few oyster shells; animal remains, chiefly cervine, human skulls. Towards the sea this bed gives place to silt (No. 7) ...	1ft.	6in.
c.	Tin ground, rounded tin ore, unmixed, and in a quartz and capel (quartz and schorl) matrix, from a few inches to 12 feet in thickness: averaging	4ft.	0in.
	Rounded pieces of slate, granite, and quartz, mixed with the tin stones.		

Mr. Henwood observes that above Carnon Section, either the old forest never flourished, or it has been destroyed in the accumulation of alluvia, in which periods of peat growth and transport of vegetable matter are indicated.

Mr. E. Smith gives a section of Carnon Works in 1807 (Geol. Trans. vol. iv. p. 404).

7. Sections given by Mr Henwood (J R Inst Corn vol. iv. pp 200, 201) which, from similarity of names, seem to refer to localities lying between Falmouth and Helston.

A. 1 The Upper part of Carn Wartha.

1. Worn and unworn granitic detritus mixed with lumps of peat, and refuse of previous operations ...	12ft	0in.
2 Tin ground—granitic sand and gravel, sprinkled here and there with waterworn granules of tin ore, interspersed at intervals with blocks of granite and schorl rock ...	12ft	0in.
Shelf of disintegrated granite		

B. At Lezerea in Mean Vroaz.

1. Peat, with nuts and branches of hazel in deeper parts, in places ...	4ft	0in.
2 Coarse granitic gravel with occasional subangular stones of tin ore ...	2ft. to 3ft	0in
3 Granitic sand, slightly mixed at intervals with felspathic clay ...	2ft	0in
4 Tin ground, angular and subangular masses of granite and schorl rock, largely mixed with tin ore of different character from that at Carn Wartha ...	3ft.	0in.

"In other parts of the Moor sections of ancient works show beds of detrital matter resting immediately on the outcrop of tin-bearing veins in the granite"

C Near Tregedna in Mawnan (? at mouth of R Helford) vegetable mould and hardened silt, 20 or 30 feet thick, overlie a poor deposit of tin ore resting on slate shelf.

(*Ibid*) Waterworn granules of pure gold have been found in detrital tin ore (which is less rounded than in other parts of Cornwall) near Helston.

Mr Henwood (T R G S Corn vol v. p 129) said that the valleys between Breagne Church and Porthleven, and from Helston to the Loo Pool, have been streamed for tin.

Penzance District.

8 *A* Mr Henwood (*op cit* p 34) gives a section in the valley between Huel Darlington and Marazion Mine near Newtown, at 20 to 30 feet above the sea. Sea sand with shells was found on vegetable matter, with trunks and branches of oak, willow, hazel in abundance, resting on poor tin ground on shelf at about the level of the sea.

B At Tregilsoe (Tregillиow), on the confines of Ludgvan and St. Hilary, a section of the short shallow vale terminating in Marazion Marsh is given by Mr Henwood (Journ R Inst Corn. vol. iv p 197). Peat about 6 feet in thickness rests on the tin ground, divided through its entire width by a thin seam of clay, impervious to water, and running obliquely both to the shelf and to the surface. Above the clay seam, the gravel consists of angular and subangular masses of slate, quartz, veinstones, granules of crystalline tin ore, all imbedded in bluish clay. Below the clay seam slate pebbles still prevail, elvan nodules are not uncommon, but the quartz is smaller and less frequent. Tin ore is diffused through the tough reddish-brown clay matrix. Although within a mile of granite no trace of granitic matter was found in these works.

Land's End District

9 *A* (Henwood, *op. cit* p 195). Near Bejowans, in Sancreed, section of a confluent with the little vale from Tregonebris to the coast at Lamorna.

1 Granitic sand and gravel with small angular and sub-angular stones	6ft to 12ft	0in
2 Peat with nuts, branches, and roots of hazel	2ft to 8ft.	0in.
3 A few inches of granitic sand, gravel, and pebbles, with occasional large granite boulders like the tin ground.		
4 Tin ground, rounded masses of felspathic granite and tin ore, fragments of veinstones and quartz crystals	2ft to 9ft	0in.

B Mr. Henwood (*op. cit* p 193) mentions the sprinkling of tin ore on S E of St. Just, in the southern and central parts of a ravine trending from Kelynack north-westward to Pornanvon. He gives a section at Boswoilas, in a narrow strip of virgin tin ground

1 Vegetable mould in some parts of the glen succeeded by	2ft or 3ft
2 Granitic gravel, sprinkled sometimes with tin ore	a few inches
3 Tin ground of granitic matter, subangular and rounded tin-bearing veinstones, pure tin stone, subangular or angular	3in to 2ft 6in

The surface of the tin ground maintains a tolerably uniform seaward slope throughout the ravine.

C. (*op cit* p 196). Between Towednack Church and Amellibrea, in the lower part of Cold Harbour Moor

1 Peat	2ft 6in.
2 Granite detritus, mixed with blue clay, and unproductive in the upper part, buff and reddish brown, with a little tin ore and tin bearing veinstones in the lower part	3ft 0ft.

D. On Leswhidden and Bostrase Moors, Mr Carne (T R G S. Corn vol iii p. 332) mentioned the occurrence of alluvial soil 6 to 9 feet in thickness on the shelf, and at Numphra Moor not exceeding 5 feet.

10 Mr Henwood (J R Inst Corn vol iv p 199) gives a section of the bed of a rivulet at St Erth, near Hayle, as follows, the thickness of the deposits not being given: Gravel, sand, and mud, on peat, under which roots, trunks, and branches of trees, with quantities of mud, were found resting on tin ground, poor and not extensive.

Mr. Carne (T R. G S Corn vol iv. pp 105–111) gives the following general notes on Diluvial tin—Cap shelves are tabular masses of rock projecting from sides or bottom of the tin ground, so as to allow of the occurrence of tin ore under them—Copper, not found in tin gravels, probably because rarely so near the surface as tin, and in the form of sulphuret so liable to decomposition—The traces of gold met with were probably derived from undiscovered veins on the east—All the productive streams occupy valleys opening on the S. coast, whilst most of the richest tin veins are near the N coast The direction of the tin streams seems to have been from N N.W. to S.S E—In narrow valleys little tin ore is obtainable In steep valleys all the ore is upon the shelf. In very gently sloping valleys tin ore is met with to within two or three feet of the surface, as at Chyanhall In gently sloping valleys the tin ground is thick but poor, owing to admixture with alluvial sediments

General Notes

As the stream tin gravels were deposited during the last stages in the elaboration of the present drainage system, their watershed boundary can scarcely have differed much from the present, it is, therefore, only natural that, whilst the richest tin veins are near the north coast the most productive streams occupy valleys opening on the south coast.

The position of the tin ground with reference to the sea-level in the estuarine sections is, unfortunately, seldom given. In Mr. Henwood's section on Marazion Green (8 *A*), mention of overlying alluvia seems to have been omitted; as Mr. Carne, in a section at Huel Darlington near Marazion River, gives twelve feet of peat and gravel above the sea sand, and the surface is given in Mr. Henwood's section at twenty to thirty feet above the sea-level, the top of the marine bed would appear to be a few feet above high water (Carne, T. R. G. S. Corn. vol. vi. p. 230).

Again, in Mr. Smith's section (5 *C*) of Lower Pentuan, the shelf is said to be at low-water level, which would place the top of the upper marine bed at about forty feet above low water, which, considering the absence of marine deposits at Wheal Virgin Works (5 *B*) and Upper Pentuan (5 *D*), is out of the question; so that either the thicknesses are not given in feet and inches, or the level of the shelf is erroneous.

Mr. Carne (T. R. G. S. Corn. vol. iv. p. 47) describes the tin ground of Drift Moor Works, near Newlyn, as resting on the sides (which come to within a few feet of the surface) and bottom (forty feet from the surface) of a clay-lined basin. This is a most exceptional phenomenon, and seems to show the great erosive power of the stream tin floods rushing into and deepening a depression, very much in the manner in which giants' kettles are produced by the pestle-like friction of fragments swirled round hollows by subglacial streams. A somewhat analogous phenomenon is mentioned by Dr. Boase, which, although not relating to stream tin, I give here (T. R. G. S. Corn. vol. iii. p. 131). "A person surveying the Channel took his station on Wolf Rock, where he observed a cavity resembling a brewer's copper, and containing rubbish at the bottom; it was covered by the sea nine hours out of twelve."

The occurrence of an oblique clay seam in the tin ground at Tregilsoe (8 *B*), separating accumulations of slightly different characters, suggests the existence of bedding, true or false. The exceptional occurrence of clay shelf (4 *G* and perhaps 5 *D*) is worthy of note.

The changeable character of the deposits in stream tin sections precludes the absolute correlation of individual beds. Inland streams cannot be expected to furnish such sections as their estuaries, yet it is scarcely safe to identify tin ground, when not overlain by sediments (as 9 *C*); when composed of fine material under a thin covering of sediment with no indication of a land surface (as in 4 *G*, *H*,

I, *J*, and 9*B*), or where it rests on outcropping tin veins (as 7 *B*), with the stanniferous gravels of Par (3), Pentuan (5 *A*, *B*, *C*, *D*), Carnon, etc. (6 *A*, *B*, *C*); whilst in some sections stanniferous deposits occur at different horizons, as 4 *E* (probably 5 *D*), 7 *B*.

To synchronize the forest remains in the various sections is unsafe, because in many valleys deposition seems to have gone on continuously, or to have been interrupted by such very brief periods of peat accumulation or undergrowth, that their relics became entirely mixed up and incorporated with the succeeding deposits, as in 4 *B*, *C*, *D*, *E*, and 5 *F*, also 4 *J* and 7 *A*.

The deposition of stream tin gravels evidently extended over a much longer period than is represented by the tin ground; for the very irregular wear of the sides and bottoms of their channels, and the existence of false shelf (4 *D*, *E*) here and there, and of masses of the surrounding rock, the apparent debris of fallen cap or false shelves (4 *A*, 5 *A*, 6 *C*), can only be accounted for by powerful streams carrying their detritus to lower levels, and occupying the energies of their upper and more torrential reaches in eroding their banks and beds into such irregular shapes as the unequal durability of the rocks permitted.

In like manner, the duration of the forest growth is not to be measured by the forest beds overlying stream tin in Marazion Marsh, Pentuan (5 *A*, *C*), etc., which can only be regarded as synchronous with a comparatively short part of the period, whilst the recurrence of peat beds with arboreal remains at different horizons in the stream tin sections (4 *A*, *B*, 5 *E*, 7 *B*, 9 *A*, 10) shows that even after the forests fringing the coasts were submerged and buried with the peat, which had accumulated around them during the last stages of their existence, it was some time before forestial growth in inland districts succumbed to unfavourable climatal conditions, and still longer before the succeeding undergrowth gave place to the bare and shrubless character presented by so large a part of western and central Cornwall now.

Although it seems only reasonable to regard the deposition of metallic detritus, as now going on, wherever the stream channels are traversed by tin veins, this process is so insignificant that as a whole the stanniferous gravels must be referred to a period considerably posterior to the raised beach formation, and, either long after the culmination of the elevation during which Head was accumulated, or in part synchronous with its accumulation, when, through greater elevation and increased rainfall, the force and volume of the streams was greater. The commencement of the forest growth is also indefinite, but subsequent to the accumulation of the Head, during the prevalence of a subsidence which produced conditions unfavourable to the existence of the tin floods as they became more suitable for its extension. So that the forest growth may have begun before the stream tin floods dwindled away, and the latter may have been partly contemporaneous with the Head. Whilst marine sediments on the forest bed or tin ground in estuarine sections (3, 5 *A*, *C*, *E*, 6 *A*, *B*, *C*, 8 *A*) prove the last great movement to have been one of subsidence, the more orderly arrangement of the deposits, the general absence of heavier far-borne detritus, the entire desertion of parts of their old channels by some of the present streams, indicate the gradual prevalence of conditions more akin to those now prevailing than to those in operation during the deposition of the stanniferous gravels.

The growth of trees, some very old, on the surface (5 *C, D*), shows that the latest of these changes must have been some time in operation, whilst the presence of human remains at great depths beneath the surface, at Carnon and Pentuan, and the tradition respecting St. Michael's Mount, would seem to justify the belief that the period in which the forests were finally submerged, although geologically very recent, is yet prehistoric.

As the subsiding movement gradually enabled the sea to circumscribe the forest tracts on its old fore-shore, the beach materials pushed forward would finally tend to bar the drainage of the valleys opening on the coast, and to convert the low lands into peat mosses, forming round the surviving trees till the further advance or dispersion of the beach dams permitted the sea to regain its old coast-line, entombing the forest fringes and their peaty surroundings beneath its sands. Eliminate from this all changes of level by internal movements, and explain the entombment of the forests by the lowering of level consequent on removal of gravel bars releasing the pent-up drainage, and the low district theory is presented. Without changes of level, however, it is perfectly untenable as applied to Cornwall, where the stream tin gravels indicate a greater elevation of the land (5 *B*), as at Carnon and Restronguet Creek (6 *A, B, C*), for instance, where the tin ground is more than sixty feet below the sea-level, whilst the estuarine deposits overlying the forest bed prove that the subsidence was progressive. Also, if the forests were submerged according to the low district hypothesis, they must have flourished under geographical conditions identical with the present, and yet these conditions have proved unfavourable to their growth on the present low lands.

On the other hand, it cannot be argued that the submerged forests are mere rafts of drift wood, stranded with vegetable matter borne down by rivers, and finally buried beneath the sea sands. The traces of submerged forests are too numerous and too extensive (1, 7, 8) to be thus accounted for; in several cases, moreover, the roots are said to occur *in situ* (3, 5, 11, ?6), and the elytra of beetles have been found (1, 6). Mr. Godwin-Austen (Q. J. G. S. vol. vi. p. 93, etc.) says: "It is diminished area and elevation which at present unfit the West of England to produce that growth of oak and gigantic fir which . . . seems to have clothed every portion of the region of Dartmoor, and which would still more be unfitted for it when at its lower Pleistocene level. On such low districts, however, and in a climate modified by a surrounding sea, some portion of a previous flora might have been enabled to live on." By substituting the words "at a few feet below its present" for "at its lower Pleistocene," the passage reads in accordance with my ideas.

PART V.

BLOWN SANDS AND RECENT MARINE

Notes on Blown Sands and Gravel Bars.

Proceeding round the coast from Plymouth.

1. Par. A low range of sand dunes separates the alluvial tracts from the Par sands.

2. Pentuan. A bank of coarse granitic sand, with bedding and false bedding indicated by black bands of schorlaceous material, dams off the sea from the low land at the mouth of Pentuan stream; on the landward margin of the low tract a low range of sand dunes has accumulated, apparently from the wind drift off the sand bank, the surface of the alluvium between them is strewn with similar granitic sand.

3 *a*. Falmouth. At the curve in the shore at Gyllyngvaes (Claypole, Proc. Brist. Nat. Soc. Ser. 2, vol. v. p. 35) the top of the gravel beach or bar coincides with the highest spring-tides.

b. Swan Pool is dammed by a bar of small quartz pebbles, 80 yards broad, and in the highest part 5 feet above high-water.

c. Mr. Godwin-Austen noticed (Rep. Brit. Assoc. for 1850, Trans. of Sects. p. 71) a platform of bare rock near Falmouth, occupying an intermediate position between high-water mark and the base of the adjacent raised beach, which varies from 3 to 10 feet above it.

d. Between Pennance Point and Maenporth, rock platforms occur at about the level of spring-tide high-water, the traces of raised beach in the vicinity being about 4 feet higher.

e. South of Maenporth, rock reefs and platforms were noticed at about 6 feet above ordinary high-water, the base of the adjacent raised beach being 10 to 15 feet above that level.

4. A strip of blown sand flanks the stream at Poljew; at Gunwalloe a considerable accumulation of blown sand covers high land between Castle Mount and Towan. On N.W. of Castle Mount, owing to the exposed situation, no blown sand occurs.

5. The Loo Pool is dammed by a bar of small quartz pebble gravel and coarse sand, with occasional flint and slate materials: coarse brown blown sand caps the low cliffs to the south of it.

6 *a*. Penzance.

Dr. Boase (T.R.G.S. Corn. vol. iii. p. 131) gives a section of the West Green sand bank, between Penzance and Newlyn, as follows:—

1. Granitic sand, of quartz, mica, hornblende slates with a little tin ore: quartz predominating .. .. 10 feet.
2. Gravel, of hornblende slate pebbles from 1 to 3 inches in diameter, 16 feet thick, resting on a submerged forest.

He points out the difference between the present sea sand and that forming the Green sand banks, between Marazion and Penzance and Newlyn, the former being finer, and composed of pulverized clay-slate and elvan, whilst the latter appears to have been derived from the destruction of a continuous band of granite between Mousehole and Cudden Point.

The original length of the Green (*op. cit.* vol. ii. p. 136) "was about three miles on the east and one mile on the west of Penzance, and is already much shortened. The ancient breadth is unknown." The West Green contained but two or three acres, and in no place exceeded 130 feet in width, when Dr. Boase wrote (*op. cit.* vol. iii. p. 131, etc.), whilst in Charles the Second's time it is mentioned in a letter to Mrs. Ley, of Penzance, as affording 36 acres of pasturage.

b. Mr. Edmonds (Edin. New Phil. Journ. vol. xlv. p. 113, for 1848) mentions the following facts. Seventy years ago a meadow lay outside the present sea-wall at the entrance to Newlyn; several houses and gardens stood on the seaward side of the cottages at Sandy Bank in Penzance, these extremities of the old Western Green are no longer visible.

In 1843 a sea-wall was built by the Corporation of Penzance to protect the remainder of the sand bank. Off the eastern bank numerous rocks between high- and low-water mark, below both sand banks, near Newlyn, Chyandower, and Marazion, buried beneath 4 to 5 feet of sand 40 years previous to 1818, were uncovered.

c. In the sand bank between Penzance and Marazion, near Marazion Bridge, Mr. Edmonds discovered a great number of land shells (*Helix virgata* and *Bulimus acutus*), in perfect preservation, throughout a depth of about 10 feet from the surface. In one instance, in the same locality, he observed a layer of small rounded pebbles, an inch or two in thickness, 3 feet below the surface of the sand, and more than 15 feet above the level of high-water. In the subjacent sand, for 4 or 5 feet in depth, he found numerous perfect land shells.

7 *a.* Whitesand Bay, to the North of Sennen Cove, is bounded by sand dunes, capping the low cliffs, and extending for a little distance inland, surrounded by higher ground.

b. On the north side of Cape Cornwall rock platforms are visible at about high-water mark, the traces of raised beach adjacent are about 6 feet above that level.

8. Lelant, Phillack, and Gwythian Towans.

"The Cornish word 'Towyn,' says Mr. Edwards (T.R.G.S. Corn. vol. vi. pp. 300–304), means 'a turfy down,' the word 'down' being perhaps a mere corruption of 'towyn' by the very common change of the letter *t* into *d*, and it is remarkable that the name 'Les Landes,' 'barren heaths,' given to the sandy districts on the S.W. coast of France, is almost precisely the same with 'Lelant,' the parish in the Towans where an ancient market town is said to have been buried by the sand. Hence Towans, Downs, Lelant, and Les Landes may all be regarded as synonymous."

In the same paper he characterizes the blown sands of St. Ives Bay as accumulations of comminuted shell sand nourishing a scanty growth of *Arundo arenaria*.

a. North of Hayle and west of Phillack an excavation of about 30 feet, at the termination of a tramway, afforded me a good section of the blown sands, here consisting of rather fine buff sand, made up of a mixture of quartz grains with comminuted shells, intersected by numerous dark bands near the top, apparently dipping northward at 10°, as though caused by the successive entombment of rank grass surfaces under gradually accumulating sand. Below the dark bands the sand still presents an appearance of bedding, such as might be occasioned by successive slips from an eminence, wherever the slopes became too sharp for the accumulating sand to rest. From this bedded appearance, and from the frequent linear distribution of perfectly preserved land shells, (*b.*) Mr. Edmonds (*op. cit.*) considered that the sand in its gradual accumulation had buried the latter "without ever completely covering the growing turf whereon the animals were feeding or hybernating."

c. Mr. Boase (T.R.G.S. Corn. vol. ii. p. 143) says, "In some places where the sand has been bored to a great depth, distinct strata separated by a vegetable crust are visible; which seem to indicate a succession of inundations at distant periods; but it is possible . . . that this may be owing to a local shifting of the sands, because in other places the like series of strata is not found."

d. In a deep cutting in the sand, about a mile from the sea, Mr. Edmonds discovered a nest of small land shells, 50 feet from the surface, of the following species:—*Helix virgata, Zonites radiatulus, Bulimus acutus, H. pulchella, Zua lubrica, Vertigo edentula, Pupa marginata, P. umbilicata, P. anglica, Bithinia ventricosa.*

He gives the following list (T.R.G.S. Corn. vol. vii. p. 71) of shells found under the surface of Phillack Towans (those marked with an asterisk are now living within 10 miles of Penzance).

Bulimus acutus.	*Helix fulva.* *	*Vertigo edentula.*
—— *obscurus.*	—— *fusca.*	—— *palustris.* *
Carychium minimum.	—— *hortensis.*	—— *pygmæa.* *
Clausilia biplicata.	—— *nemoralis.*	*Vitrina pellucida.*
Conovulus bidentatus.	—— *pulchella.*	*Zonites alliarius.*
—— *denticulatus.*	—— *virgata.*	—— *cellarius.*
Helix aspersa.	*Pupa anglica.*	—— *nitidulus.*
—— *caperata.*	—— *marginata.* *	—— *pygmæus.*
—— *ericetorum.*	—— *umbilicata.*	—— *rotundatus.*

Mr. Edmonds mentions the occurrences of numerous shells of *Helix pulchella*, at depths varying from 1 to 30 feet, in various parts of the sands, and says that living specimens have been observed, and that their exuviæ have been found in Whitesand Bay sandhills as well as those near Gunwalloe and Mullion, Mounts Bay and Gorran (on the South Coast of East Cornwall).

Mr. Crouch, who identified the species given above, observes that *Helix pulchella* is uncommon in the locality, that it has been found by him near Falmouth, at Pendennis, and near Penzance, at Trereife; also near the Land's End.

From the quantity of shells found in so small a space in the Towans, Mr. Crouch considers that they were once abundant in Cornwall, but are now gradually becoming extinct.

Pupa marginata and *Bithinia ventricosa* he alludes to as rare, a few dead shells having been obtained by him at Whitesand Bay (Land's

End) and near Hayle, but that no live specimens have been found in Cornwall

e Near Godrevy Island, rock platforms are visible at about the level of spring-tide high-water, the base of the adjacent raised beach is from 4 to 5 feet above ordinary high-water

9 Mr N Whitley (25th Ann. Rep Roy Inst. Corn for 1843) mentions "the succession of sand hills, principally composed of comminuted shells, covering about 1,500 acres, on the north-east of Perran Porth. The inland portion," he says, "being level and well sheltered, might easily and profitably be reclaimed by an admixture of clay with the sandy wastes, as in Norfolk, where by this means a free sandy loam, forming a most productive soil, has been obtained Owing to the extent of the Perran Sands, being more heated by the sun's rays than the surrounding districts, in calm weather by the radiation of heat from the sand hills, it is often oppressively warm at the Porth during the early part of the night"

10 The patch of blown sands bordering Hollywell Bay may be regarded as a continuation of the Perran Sands, it is partly bounded by a stream

11 The flattish tract between New Quay and Fistral Bay is covered by blown sand

12 Sand dunes occur at Porth Barn, Mawgan Porth, and Porthcothan, Tregarnon, and Permizen bays, they are very insignificant.

13. Between Constantine and Perleze Bays a low tract is covered by blown sand, as exposed near Constantine Island (*vide* Raised Beaches, 19 *e*), it is 4½ feet in thickness, and contains layers of *Patellæ* and broken *Mytili*, and occasional angular slate fragments at the base.

14 The low tract in which St. Enodock's Church is situated is composed of blown sand.

15 In Perleze Bay, and near Port Isaac, rock platforms were noticed at about ordinary high-water mark.

General Notes.

Wherever the area covered by the blown sands is extensive, we note that the lands generally lie low with reference to the sea or relatively to the surrounding country That the accumulation spreads from west to east, and only occurs in considerable quantity in localities at or near the coast-line facing westwards

Thus, in bays where the cliffs are very low and unbroken by gorges or stream channels, facing westwards and receiving the full force of winds and waves of the Atlantic, the most favourable conditions occur for æolian transport on the Cornish coast

Naturally, the inland extension of the sand depends upon the extent of low-lying country; but, besides this check on its extension exercised by barrier hills, running water and the growth of certain plants may arrest its progress, the former intercepts the fugitive grains which seldom rise more than a few inches above the ground and are suspended for a short time (De la Beche, Report, etc, p. 446) As to the latter, Major T Austin (Proc Brist Nat. Soc., vol ii. No. 11, for Dec 1867) gives the following plants as best suited to arrest the inroads of blowing sand, in some cases by collecting hillocks kept together by their matted roots—*Ammophila arenaria* (sea reed); *Triticum junceum* (sea wheat grass); *Hippophæ rhamnoides* (sand thorn), *Cakile maritima* (sea rocket); *Salsola kali* (salt wort), and *Sonchus* (sand thistle).

Mr. Henwood (40th Ann. Rep. Roy. Inst. Corn. for 1858) alludes to the progress of the sand drift covering the low lands of St. Minver on the east of Padstow, being checked by the growth of *Arundo arenaria*.

The appearances of bedding in the blown sands are worthy of note as they betray the incipient characters which in the old blown sands of Fistral Bay and Greenway have developed on consolidation into marked laminæ or thin flaggy sandstones, and near Godrevy and New Quay into thick beds. Although the constant shifting and accumulation of the sands (8 *b*) upon a growing surface must be true, yet the final entombment and successive growth of grass, or *Arundo arenaria*, is more likely to have been occasioned by heavy gales drifting large quantities of sand upon the dunes (8 *c*), for, constant shifting of particles would be less likely to produce definite layers; the cohesion of the particles of successive surfaces of comminuted shell sand lending itself readily to the formation of definite beds, and when counter wind drifts prevailed, to false beds, in the process of consolidation through the downward passage of rain waters. But as far as I am aware no traces of old vegetable surfaces have been found in the old consolidated blown sands. The false-bedded appearance is well shown in the old blown sands of Barnstaple Bay. The thin layers of schorlaceous and quartzose grains in the sand bank at the mouth of the Pentuan Valley seem to be due to marine action, sorting the materials.

The absence of sand or gravel bars on parts of the Cornish coast directly exposed to the waves of the Atlantic, and their limitation on the southern coast to sites where promontories and headlands shelter them from the direct influence of the prevalent winds, and where the rapid transport of shingle is lessened by projections of the coast on the further side, is worthy of note. Thus, the West Green bank sheltered by the Land's End district occurs in the centre of Mounts Bay, the Loo Bar, somewhat similarly sheltered, has been piled up where the southerly trend of the Lizard coast-line becomes pronounced; the Swan Pool Bar and the extensive beaches of Falmouth, lying between the flow of the Fal and Helford nearly at right angles, are sheltered in a measure by the Lizard district, and the further transport of shingle is checked by the projection of Pendennis Point.

The set of the coast-line has been aided by the inability of the stream waters to keep a seaward passage clear, as in the case of the Loo Pool, which represents the ponded drainage of the Cober and its tributaries. The ceremony of cutting the Bar annually to allow the waters to escape more rapidly than by filtration through it, and thus prevent floods, shows how effectually the seaward outlet of the stream has been overcome. The finer accretions to some of the banks, as in the West Green, have been shifted higher by winds, a tongue of sand occurs on the east of the Loo Pool similarly drifted

The surfaces of the planed Killas reefs, of which I have only given a few examples, occupy in most cases a position intermediate between the base of the several raised beaches in their vicinity and high-water mark (3 *c*, *d*, *e*; 7 *b*, 8 *e*, 15) Mr Godwin-Austen attributed (Rep Brit Assoc for 1850, Trans of Sects p 71) their positions to a recent elevation (preceded by a subsidence) of not more than 10 feet In further proof of this he cites the mud beds of the Exe and Sussex Ouse, containing estuarine shells at slight elevations above the present sea-level. The occurrence of many of the rock platforms are explainable without invoking changes of level The comparatively recent subsidence by which the forest lands were submerged would have brought again within the influence of the waves such portions of the old platforms, upon which the raised beach rested, as had survived the intervening subaerial waste, and, whilst robbing them of whatever superimposed deposits might have existed, would plane anew those more durable portions which came within the influence of the waves, leaving others shorn of their deposits, marking by the heights of their surfaces the seaward slope of the old plane of marine denudation. Bearing in mind the very unequal heights of old beaches of the same age, and the irregular levels of their platforms in places at the base of the same cliff (in places, as in Fistral Bay, the base of the raised beach occupies an almost uniformly persistent level), except where great discrepancies in their levels with reference to adjacent raised beaches occurred, the platforms might be explained as above. Other phenomena, however, whilst in no way interfering with the above explanation, would appear to favour the idea that a pause in the downward movement, after the submergence of the forests, was succeeded by a slight contrary movement Such an oscillation might serve to explain the river sediments gaining on the marine, in estuarine stream tin sections, and to enable them to continue *pari passu* with a resumption of the subsiding movement If, from the sections in Marazion Marsh given by Messrs. Henwood (T R G S Corn vol v p 34) and Carne (*Ibid* vol vi p 230, etc), we may place the top of the marine bed at 2 or 3 feet above high-water, an oscillation would alone account for its position. The formation of the Warren Sand Bank and Northam Pebble Ridge might also be explained by a slight elevation, whilst the rapid diminution of both would seem to indicate a return to the previous contrary movement The formation and diminution of the West Green Sand Bank might be similarly explained

Mr. Edmonds (Edin New Phil Journ for 1848), commenting on the diminution of the bank, says, that 300 years ago, in Leland's time, the causeway leading to St. Michael's Mount was uncovered six hours out of twelve, and continued so for 220 years The passage to the Mount in 1848 was open four hours out of twelve, and often during strong S W winds covered at neap tides for days together He ascribed these rapid changes (6 *b*) within 80 years to the removal of sand, which supported the western side of the ridge, for ballast and agricultural purposes. "Some idea," he says (T. R. G. S. Corn vol. vii. p 31), "of the vast quantity of sand thus abstracted (for manure) may be formed by the fact that a very usual clause in farming leases in this neighbourhood is, 'That ten butt loads of sea sand shall be spread on every acre whenever it is broken for tillage.'" This explanation is a very plausible one, and, coupled with the hypothesis before mentioned, would be a powerful adjunct in accounting for more rapid recent waste. Great quantities of comminuted shell sand are also carted from Bude by the farmers of Northwest Devon

In conclusion, I have to express my sincere thanks to Mr. W. Whitaker, Dr. C Le Neve Foster, and to Mr. E Parfitt, of Exeter, for kindly furnishing me with all the information in their power concerning the literature of the subject, to Mr Robert Hunt, F.R.S., Keeper of the Mining Records, for placing at my disposal some beautifully executed sections of the St Agnes deposits by Mr A C Davies, some of which I have submitted to the Geological Society in a reduced form; also to Mr Horace B. Woodward for the kind interest he took in this paper in its original form. and the information he obtained for me as to the best means of insuring its publication

Stephen Austin and Sons, Printers, Hertford

Milton Keynes UK
Ingram Content Group UK Ltd.
UKHW022014120124
435957UK00005B/113